PRE AND PROBIOTICS FOR POULTRY GUT HEALTH

FSC
www.fsc.org
MIX
Paper | Supporting responsible forestry
FSC® C022174

Poultry Science Symposium Series

Executive Editor (Volumes 1–18): B.M. Freeman

1 Physiology of the Domestic Fowl*
2 Protein Utilization by Poultry*
3 Environmental Control in Poultry Production*
4 Egg Quality – a Study of the Hen's Egg*
5 The Fertility and Hatchability of the Hen's Egg*
6 i. Factors Affecting Egg Grading*
ii. Aspects of Poultry Behaviour*
7 Poultry Disease and World Economy
8 Egg Formation and Production
9 Energy Requirements of Poultry*
10 Economic Factors Affecting Egg Production*
11 Digestion in the Fowl*
12 Growth and Poultry Meat Production*
13 Avian Coccidiosis*
14 Food Intake Regulation in Poultry*
15 Meat Quality in Poultry and Game Birds
16 Avian Immunology
17 Reproductive Biology of Poultry
18 Poultry Genetics and Breeding
19 Nutrient Requirements of Poultry and Nutritional Research*
20 Egg Quality – Current Problems and Recent Advances*
21 Recent Advances in Turkey Science
22 Avian Incubation
23 Bone Biology and Skeletal Disorders
24 Poultry Immunology*
25 Poultry Meat Science
26 Poultry Feedstuffs
27 Welfare of the Laying Hen
28 Avian Gut Function in Health and Disease
29 Biology of Breeding Poultry
30 Alternative Systems for Poultry: Health, Welfare and Productivity
31 Sustainable Poultry Production in Europe
32 Poultry Feathers and Skin: The Poultry Integument in Health and Welfare
33 Pre and Probiotics for Poultry Gut Health

*Out of print
Volumes 1–24 were not published by CAB International. Those still in print may be ordered from:

Carfax Publishing Company
PO Box 25, Abingdon, Oxfordshire OX14 3UE, UK

Pre and Probiotics for Poultry Gut Health

Poultry Science Symposium Series
Volume Thirty Three

Edited by

Helen Masey O'Neill
AB Agri, Peterborough, UK

Emily Burton
Nottingham Trent University, UK

Dawn Scholey
Nottingham Trent University, UK

CABI

CABI is a trading name of CAB International

CABI
Nosworthy Way
Wallingford
Oxfordshire OX10 8DE
UK

Tel: +44 (0)1491 832111
E-mail: info@cabi.org
Website: www.cabi.org

CABI
200 Portland Street
Boston
MA 02114
USA

Tel: +1 (617)682-9015
E-mail: cabi-nao@cabi.org

ISBN-13: 9781800622722 (hardback)
9781800622739 (ePDF)
9781800622746 (ePub)

DOI: 10.1079/9781800622746.0000

Commissioning Editor: Alexandra Lainsbury
Editorial Assistant: Emma McCann
Production Editor: Rosie Hayden

Typeset by Straive, Pondicherry, India
Printed and bound in the UK by Severn, Gloucester

CONTENTS

Contributor List

S. Amir, Nottingham Trent University, UK. E-mail: Saba.amir@ntu.ac.uk
Allan Ball, Slate Hall Veterinary Practice, Lincoln, UK. E-mail: allan.ball@slatehall.co.uk
Mike R Bedford, AB Vista, Marlborough, UK. E-mail: Mike.bedford@Abvista.com
Ralph Bishop, Premier Nutrition, Rugeley, UK. E-mail: Ralph.Bishop@premiernutrition.co.uk
Emily Burton, Nottingham Trent University, UK. E-mail: Emily.burton@ntu.ac.uk
E. Chevaux, Lallemand SAS, France. E-mail: echevaux@lallemand.com
Martha R.J. Clokie, University of Leicester, UK. E-mail: mrjc1@leicester.ac.uk
Nicolae Corcionivoschi, Agri Food and Biosciences Institute, Belfast
Tom Dale, AB Vista, Marlborough, UK. E-mail: Tom.Dale@Abvista.com
V. Demey, Lallemand Animal Nutrition, France. E-mail: vdemey@lallemand.com
A. Doherty, Vetagro S.p.A., Reggio Emilia, Italy. E-mail: aidan.doherty@vetagro.com
Richard Ducatelle, Faculty of Veterinary Medicine, Ghent University, Belgium. E-mail: Richard.Ducatelle@UGent.be
Venessa Eeckhaut, Faculty of Veterinary Medicine, Ghent University, Belgium. E-mail: venessa.eeckhaut@ugent.be
C.L. Elvis-Chikwem, University of Nottingham, UK. E-mail: chinwendu.elvis-chikwem@nottingham.ac.uk
Salvatore Galgano, Monogastric Science Research Centre, SRUC, Edinburgh, UK. E-mail: Salvatore.Galgano@sruc.ac.uk
Evy Goossens, Faculty of Veterinary Medicine, Ghent University, Belgium. E-mail: evy.goossens@ugent.be
Ozan Gundogdu, London School of Hygiene and Tropical Medicine, London
Eliana Henríquez Rodriguez, Pen & Tec Consulting S.L.U., Spain. E-mail: eliana@pentec-consulting.eu
Steve P.T. Hooton, University of Leicester, UK.
Farina Khattak, Monogastric Science Research Centre, SRUC, Edinburgh, UK. E-mail: Farina.khattak@sruc.ac.uk
Roberto La Ragione, University of Surrey, UK. E-mail: r.laragione@surrey.ac.uk
Elinor McCartney, Pen & Tec Consulting S.L.U., Spain. E-mail: elinor@pentec-consulting.eu

Samantha McLean, Nottingham Trent University, Nottingham, UK. E-mail: samantha.mclean@ntu.ac.uk
A. Meuter, Chr. Hansen, Denmark. E-mail: franme@chr-hansen.com
Natalie Morgan, Curtin University, Perth, Australia. E-mail: natalie.morgan@curtin.edu.au
M. Naeem, Nottingham Trent University, UK. E-mail: Naeem.naeem@ntu.ac.uk
Daniel Parker, Slate Hall, Metheringham, Lincolnshire, UK. E-mail: Daniel.parker@slatehall.co.uk
D. Préveraud, Adisseo, Antony, France. E-mail: damien.preveraud@adisseo.com
Anne Richmond, Moy Park Ltd, Craigavon, UK. E-mail: Anne.Richmond@moypark.com
F. Riva, University of Glasgow, Scotland, UK. E-mail: francesca.riva@glasgow.ac.uk
Ivan Rychlik, Veterinary Research Institute, Brno, Czech Republic. E-mail: ivan.rychlik@vri.cz
Dawn Scholey, Nottingham Trent University, UK. E-mail: dawn.scholey@ntu.ac.uk
Reg Smith, The Stables, Princethorpe, Warwickshire, UK. E-mail: Reg@Princethorpepoultry.co.uk
Anisha M. Thanki, University of Leicester, UK. E-mail: at263@leicester.ac.uk
C. Torres, Zinpro Corporation, MN, USA. E-mail: ctorres@zinpro.com
W. Van der Veken, Huvepharma NV, Belgium. E-mail: wouter.vanderveken@huvepharma.com
Filip Van Immerseel, Faculty of Veterinary Medicine, Ghent University, Belgium. E-mail: filip.vanimmerseel@ugent.be
A. Wealleans, Kemin Europa NV, Belgium. E-mail: Alexandra.Wealleans@kemin.com

PREFACE

The Symposium, the 33rd in the Poultry Science Symposium Series, was held on 22–24 August 2022 at Clare College, Cambridge University, UK. The Symposium took place after a 2-year postponement due to the Covid-19 pandemic. The Symposium featured global experts discussing the background context, challenges and opportunities of gut health in modern broiler production. This book contains the manuscripts produced by the presenting authors and their supporting co-authors or, in some cases, a modified transcript of their talk.

The title of the Symposium was 'Pre and probiotics; nutritional, veterinary and production perspectives'. This meeting was the first directly nutrition-related symposium in the series since the 28th Symposium in 2005. The marketplace for additives and the way in which gut health is addressed has changed dramatically in the interim, not least due to the removal of antibiotic growth promoters (AGPs) in Europe in 2006. The aim was to draw together many perspectives on this topic, starting with an overview of AGPs, how they worked and the challenges that occur in their absence, thorough definitions of additives in the pre and probiotic space and examples of their mechanisms. We addressed how to test them and how they interact with other additives with learning from our own and allied sectors. The unfortunate delay owing to Covid-19 gave us an unexpected advantage of being able to include some very novel perspectives, including Professor Martha Clokie and Dr Anisha Thanki discussing the potential for bacteriophage use in poultry production.

One of the most interesting discussions during the Symposium was around whether or not, in Europe, we should still consider AGPs to be a benchmark positive control in poultry production. During the symposium, we were reminded that AGPs gave a significant gut health benefit and that there are now many non-medicinal alternatives and advances in knowledge of biosecurity. However, now that AGPs are no longer used, should we still consider an equivalent effect to be achievable? If they are no longer used, is it even relevant to use the term 'antibiotic alternative'? The overwhelming feeling from the audience was that we should not. The use of the word antibiotic in the context of nutrition distracts from the good work done in responsible medicine use, genetic development of broilers and, not least, the revolution of feed additives in animal production.

These symposia only happen because of the volunteers who graciously give their time and energy for free. Countless hours have been given by many colleagues on the organizing committee who gave their time to the Symposium and book alongside their day jobs. The scientific committee are responsible for the diverse and interesting programme, put together with care and thought and with some adversity brought about by the pandemic. There were also many other volunteers who helped with administration and practical help, and we very much appreciate everyone who has been involved. I would like to make special mention of Dr Dawn Scholey, our co-editor, who was particularly influential in the development of this book. The UK Branch of the World's Poultry Science Association (WPSA) is a thriving and friendly community and these symposia are continued evidence of this. We hope you enjoy these proceedings.

Nell Masey O'Neill
AB Agri Limited
Nottingham, UK
February 2023

33rd Poultry Science Symposium, Cambridge 2022

President's Preface

Gut health and the consequences of intestinal imbalance remain at the top of most broiler producers' concerns in recent years, in terms of health, welfare and sustainable production. This is especially relevant in a so-called 'post-antibiotic era' where alternative and effective tools and management are called for as part of antibiotic reduction strategies.

The 33rd Symposium was a welcome addition to the discussions and progress on this important topic. The organizers are to be congratulated for bringing together all the major players in this complicated and often frustrating area of poultry production in endeavouring to establish and maintain a balanced and efficient microbiome.

I am indebted to all those involved but most especially to the contributors to the broad breadth of presentations and discussion sessions. These were diverse and thought provoking and, if not solving all our problems, certainly gave us a broad view and pointers on where we are, where we have gone wrong and where we might go in the future to make even more progress.

The Poultry Science Symposium series attempts to address significant and complex current issues by drawing on a broad mix of experience and expertise. The 33rd symposium epitomized this approach to the 'appliance of science' and achieved one of my goals in bringing together the best in science and innovative thinking with those working at the 'coal face' of poultry production. It is only through such collaboration that we are likely to maintain and improve poultry health, welfare and sustainable production.

Not only did the delegates participating during the event benefit from the presentations and the discussions triggered, but through the commitment of the authors and contributors another major output from the symposium has been these proceedings, which have created a full and lasting resource from which a much larger audience will benefit.

The symposium may not have come up with all the answers but has certainly helped to signpost where we might go in the future to achieve our goals.

Stephen Lister
President, WPSA UK Branch

CHAPTER 1
What is Antimicrobial Resistance and Why is it a Problem?

Samantha McLean*

School of Science and Technology, Nottingham Trent University, Nottingham, UK

Abstract

Antimicrobial resistance (AMR), which occurs when previously susceptible bacteria become resistant to the drug(s) used to treat them, is a major global threat to animals, ecosystems and human health. With rising rates of AMR greatly outpacing the development of new antibiotics and antimicrobial drugs, resistant strains are increasingly identified within the community and environment where previously they were largely restricted to the hospital setting. Addressing the public health threat posed by AMR is a national strategic priority for many nations including the UK. However, resistance is a complex issue driven by many interconnected factors. It is an issue that unites public health, animal welfare, food safety and food security, with the food chain being regarded as an important transmission route for resistant bacteria to enter humans. Antibiotic resistance occurs naturally but is accelerated by the misuse and overuse of antibiotics, as well as poor infection prevention and control. In areas where antibiotics can be bought for human or animal use without a prescription, the emergence and spread of resistance is exacerbated. Similarly, in countries without standard treatment guidelines, antibiotics are often over-prescribed by health workers and veterinarians and over-used by the public. Therefore, steps must be taken at all levels globally to limit the spread of resistance. With annual death rates attributed to AMR predicted to reach ten million by 2050 unless urgent global action is taken, we could be heading towards a post-antibiotic era, where common infections and minor injuries can once again become fatal.

*Corresponding author: samantha.mclean@ntu.ac.uk

DOI: 10.1079/9781800622746.0001

1.1 Introduction

Antibiotics, once viewed as the golden bullet in the fight against bacterial infections, now increasingly fail owing to the rise of antibiotic resistance. The problem of antibiotic resistance and the wider AMR is not new; penicillin was first introduced commercially in the early 1940s but resistance had already been identified by 1940 (Rammelkamp and Maxon, 1942; Abraham and Chain, 1988). By the 1960s, around 80% of *Staphylococcus aureus* infections were found to be penicillin resistant irrespective of their origin (Lowy, 2003). This trend of resistance has been repeated after the introduction of every antibiotic to date.

Antibiotics are medicines, often derived from fungi and some bacterial species or synthesized in a laboratory, that are absorbed by the body and can kill bacteria or inhibit their growth allowing the host immune system to remove the invading pathogens. The term antimicrobials refers to a wider group of drugs that includes antibiotics but also other antimicrobial compounds such as those able to inhibit or kill fungi, parasites and viruses.

There are two main types of AMR, known as intrinsic resistance and acquired resistance. Intrinsic resistance generally refers to whole species of organisms, rather than single isolates or strains, because it is the natural characteristics of the species that causes the resistance to a given antibiotic and does not require the acquisition of genetic material nor mutation to occur. For example, an organism could naturally be impermeable to a drug due to the structure of its cell wall, or intrinsically resistant to an antimicrobial because it does not possess the target of that the drug. Intrinsic resistance has been redefined by the European Committee on Antimicrobial Susceptibility as the 'expected resistance phenotype' as this type of resistance can often be predicted, whereas true resistance is acquired by a susceptible population (Gatermann *et al.*, 2022). In contrast to this expected resistance phenotype, microbes can develop acquired resistance. Here, previously susceptible populations of microbes acquire resistance to an antimicrobial agent over time, often in response to exposure. Acquired resistance has significantly reduced the therapeutic value of many important antibiotics, as is the case for the first commercially produced antibiotic: penicillin. However, not all resistance is acquired at the same rate; whilst resistance to penicillin emerged in *S. aureus* around the time of its introduction, *Streptococcus pyogenes* remains largely susceptible to penicillin, although there is no guarantee that this will remain the case in the future. There have been sporadic reports on *S. pyogenes* non-susceptibility and resistance to penicillin emerging in China (Yu *et al.*, 2020). There is no clear explanation for the rate or extent of acquisition of resistance between different species, which is probably due to a multitude of interconnecting factors including genome plasticity and the microbial community where the organisms reside (Patel, 2016; Bottery *et al.*, 2021). The genetic capacity for resistance does not always explain the varied rates in the emergence of resistance and whilst selective pressure can lead to a rise in antibiotic resistance this is not always the case.

Acquired resistance occurs in one of two ways. First, a single cell may mutate and that mutation could confer resistance to an antimicrobial. The second way in which micro-organisms acquire resistance is via the transfer of genetic material that contains resistance genes and primarily this occurs on mobile genetic elements such as plasmids (Blair *et al.*, 2015). It is important to remember that mutations within a bacterial population occur by chance and only some of these mutations give rise to AMR. There are numerous factors that drive the evolution of bacterial populations that can result in the emergence of resistance, including interactions with differing environments and molecules in the human or animal host, competition with the host commensal microbial population and exposure to drugs including antibiotics (Bliven and Maurelli, 2016). Undoubtedly, exposure to antibiotics drives the emergence of resistant groups as the dominant population due to the susceptible population being inhibited or killed by the antimicrobial, reducing competition for the resistant cells. In the absence of the antibiotic, these resistant bacterial cells may never have become the dominant population because often resistance to antibiotics confers a fitness cost to the bacterial cells (Melnyk *et al.*, 2014). However, owing to the selective pressure exerted by the presence of the antibiotic, the susceptible population of bacterial cells die out, leaving the resistant population to become the dominant population. Bacterial populations may also encounter sub-lethal concentrations of antibiotics in the natural environment or in animal and human hosts due to non-compliance or poor drug pharmacokinetics; here resistance can emerge that is less likely to carry a fitness cost to the cells such as altered metabolism, changes in gene regulation and biofilm formation, which may lead to altered fitness effects of subsequent drug-resistant mutations (Matange *et al.*, 2019).

Bacterial resistance traits are passed directly from parent to daughter cells via vertical transfer during replication. They are also often passed directly from one bacterial cell to another via a mechanism of horizontal gene transfer, spreading the antibiotic resistance phenotype throughout a population. There are multiple mechanisms for the spread of resistance by horizontal gene transfer including via plasmids, transducing particles, outer membrane vehicles and extracellular DNA. The subject of these mechanisms of genetic transfer is comprehensively reviewed in Partridge *et al.* (2018).

The spread of a resistant population of microbes throughout an environment can be achieved via multiple pathways. Resistance regularly emerges within animal and human populations, with the extent to which the environment plays a role in the emergence of resistance currently under study. However, the expectation is that the environment will play some role in emergence and a significant role in the transmission of resistant populations (European Commission, 2017; Bengtsson-Palme *et al.*, 2021). The transmission of resistant microbial populations between humans, animals and the environment occurs via a complex interconnecting network of routes. As humans, companion animals and food-producing animals rely on similar types of antimicrobials to treat infectious disease, resistance can emerge via all these routes. From there, resistant microbial populations can disseminate into the other groups and into the environment. Resistant microbes can pass from human to human, animal to animal, or between species via close contact. However, they can also spread through the food chain, via the production of meat from food-producing animals and via the environment due to the use of fertilizers on crops that are contaminated with antimicrobial-resistant microbes. Resistant microbes can also be released into the environment via wastewater treatment facilities that discharge into waterways; although these facilities may reduce the microbial burden of the discharged water, they are unable to eradicate microbes entirely. Additionally, antibiotics themselves can enter waterways via wastewater-treatment facilities because there is no effective mechanism to remove all antibiotics during sewage treatment. Antibiotics can also reach waterways via discharge from antibiotic manufacturing sites either directly into waterways or via discharge to wastewater-treatment facilities (Li *et al.*, 2008; Kristiansson *et al.*, 2011; Karkman *et al.*, 2019).

This chapter provides readers with an understanding of the global scale of the AMR threat that we face, alongside some specific mechanisms and considerations, concluding with a review of the antibiotic stewardship principles and how they can be employed to reduce the spread of this deadly resistance.

1.2 The Global Scale of Antimicrobial Resistance

To understand the scale of the challenge that we face in addressing AMR, it is important to first understand the history of antibiotic use. The emergence of antibiotic resistance after the introduction of antibiotics to market is not predicable, with some strains of bacteria becoming resistant to antibiotics very soon after their commercial introduction and others taking much longer to acquire resistance. Similarly, some antibiotics are used for many years before resistance emerges in any microbial population and for other antibiotics this can occur within the same year of commercial introduction. The one consistent factor in the use of antibiotics is that, after the introduction of an antibiotic for treatment, resistance seems inevitable. The numbers of resistant infections reported and subsequent deaths resulting from AMR are rising annually with the latest comprehensive estimate stating that in 2019 alone 1.27 million deaths were directly attributable to AMR (Murray *et al.*, 2022).

The concept of antibiotic resistance is not new. In his Nobel lecture in 1945 just after the commercial introduction of penicillin, Sir Alexander Fleming warned of the dangers of misuse and overuse of this antibiotic stating that 'the time may come when penicillin can be bought by anyone in the shops. Then there is the danger that the ignorant man may easily underdose himself and by exposing his microbes to non-lethal quantities of the drug make them resistant' (Fleming, 1945). The first treatment of a patient using penicillin produced under the Office of Scientific Research and Development was in March 1942. Thereafter production skyrocketed to 21 billion units in the USA alone by 1943 and to more than 6.8 trillion units by 1945 (Aldridge and Sturchi, 1999). The discovery and subsequent commercial production of penicillin and other antimicrobials such as the sulfa drugs led to a golden era of antimicrobial drug discovery with many new classes being discovered, developed and marketed (Hutchings *et al.*, 2019). Their use was not limited to the treatment of infection in humans and animals; the growth-promoter effects of antibiotics on animals was discovered in the 1940s, where long-term sub-therapeutic administration of antibiotics to animals was found to promote growth and reduce susceptibility to infection. The use of antibiotics was approved by the Food and Drug Administration (FDA) in the USA without veterinary prescription as early as 1951, with European countries soon following (Castanon, 2007). Antibiotic use in the care of food-producing animals became so widespread that antibiotic consumption by animals far outweighed their use in human medicine. The current picture of antibiotic use in food-producing animals globally is complex, with some countries including the UK and European Union (EU) members imposing ever tightening regulation on the administration of antibiotics to food-producing animals for anything other than the therapeutic treatment of an identified infection and the EU banning the use of antibiotics as growth promoters in 2006. This led to antibiotic use in food-producing animals dropping below human medicinal use. However, these tight EU regulations are not reflected in worldwide practice, which varies considerably between countries. The latest data show that 44% more medicinally important antimicrobial drugs are consumed by animals than by humans in the USA (Wallinga and Kar, 2020; EFSA, 2021).

Global usage practices of antibiotics and the concurrent rise in antibiotic resistance are of major concern due to the massive impact that antibiotics have had on our lives. For example, in the pre-antibiotic era it is estimated that five out of every 1000 women who gave birth died owing to infection, three out of ten people who contracted pneumonia died and one in every ten people who suffered a skin infection from a scrape, cut or bite lost a limb (McKenna, 2013). Whilst modern medicine has advanced significantly since the first commercial use of antibiotics for widespread treatment of infections, the loss of these life-saving drugs would increase the risks associated with numerous modern treatments and procedures, including cancer treatments, caesarean sections and many other types of surgery.

1.3 The Fragile and Failing Antibiotic Pipeline

Given the historical success of the antibiotic production pipeline, it would seem logical that the answer to this resistance crisis is to develop new novel antibiotics, to keep one step ahead of the evolving pathogens. However, the emergence of resistance is accompanied by a fragile and failing antibiotic pipeline, as highlighted recently in a report published by the Biotechnology Innovation Organization (Thomas and Wessel, 2022). The report on antibacterial innovation concluded that 'the breadth and novelty of the clinical antibiotic pipeline is insufficient to meet the growing threat of antibiotic resistant pathogens'. Many of the antibiotics that have been commercially introduced in recent years were discovered many years previously; for example, the lipopeptide antibiotic daptomycin was discovered in 1983 but, owing to concerns over its safety, it was not approved for clinical use until 2003 (Ledger *et al.*, 2022).

As of early 2022, there were only 44 drugs in the antibiotic pipeline (excluding those used for the treatment of tuberculosis and *Clostridium difficile* infections), which, based on historic success rates for bringing drugs in the pipeline to market, is far below the number required to address the threat posed by the emergence of increasingly resistant bacterial pathogens (Thomas and Wessel, 2022). These figures are also far below the numbers of drugs in trials for other diseases; for example, within the period 2020–2021 there were 1030 drug and vaccine programmes for treatment or prevention of COVID-19 and within the same timeframe there were 160 new chemical entities in clinical trials for breast cancer. With the expectation that if we do not take significant action to reduce the spread of AMR including bringing new antimicrobials to market that are effective against resistant pathogens, the numbers of deaths attributable to antimicrobial-resistant infections may well exceed deaths caused by all cancers by 2050. There is therefore still a significant underinvestment in the research and development of antibacterials in comparison to drugs for other diseases and this underinvestment is in stark contrast to the magnitude of the threat faced.

Evidence from the USA suggests that investment in bringing new antibiotics to market in recent years has primarily fallen to small companies (80%) and non-profit organizations and universities (8%), with large companies only accounting for 12% of the antibacterial drugs in clinical trials in 2020–2021. The reason that large pharmaceutical companies are leaving the antibacterial drug development market is the risk to investment. Novel antibiotics that successfully enter the market will be held in reserve as drugs of last resort only to be used where other antimicrobial therapies have failed. Additionally, antibiotic treatments are classified as short term and are therefore much less profitable than many other drugs on the market, particularly the so-called blockbuster drugs that generate sales worth over US$1 billion annually. The reduced interest of the large pharmaceutical companies in developing new antibiotics makes investment by other groups such venture capitalists much riskier due to reduced opportunities for licensing deals and mergers and acquisitions. This risk is highlighted further still by the bankruptcies in recent years of numerous small- and medium-sized enterprises that successfully developed and brought new antibiotics to market with FDA approval but were still unable to attract sufficient investment to remain financially viable (Klug *et al.*, 2021). Although policy and regulatory solutions, including push and pull incentives are being offered they are still yet to combine in a manner that sufficiently incentivizes large pharmaceutical companies and investors, with many reporting that there is still little to no investment return for antibiotics. It is generally accepted that to stabilize the antibiotic development pipeline and bring new antibacterial drugs to market that will impact the fight against AMR, solutions from governments and regulatory bodies need to span the entire drug discovery and development pathway with early-stage investment in research, later stage investment in development, clear and efficient regulatory pathways and post-approval market incentives (Thomas and Wessel, 2022).

1.4 What are the Consequences of Entering a Post-antibiotics Era?

Given the lack of novel antibiotic discovery, coupled with the rise in antibiotic resistance observed globally, we currently face the very real possibility of a post-antibiotic era, where antibiotic treatments become ineffective. The first of a series of publications, commissioned by the UK government and written by economist Professor Jim O'Neill, used detailed scenario analyses of data including the annual deaths due to antimicrobial-resistant infections to produce an estimation of the deaths that will be attributable to AMR by 2050 if no significant change in practice occurs. They estimated that at the time of publication 700,000 deaths per annum were directly attributable to AMR globally. However, the scenario analyses estimated that deaths directly attributable to AMR could skyrocket to ten million people worldwide by 2050 if we do not take action to prevent the spread of AMR (O'Neill, 2014).

If AMR pushes us to the brink of a post-antibiotic era, where antibiotics are effectively useless in treating most bacterial infections, the implications would not only be felt by those incurring an infection as their primary ailment. The wider impact of losing the efficacy of antibiotics would be staggering, antibiotic prophylaxis and treatment is essential to many modern medical procedures and enables critical interventions including Caesarean sections, chemotherapy, organ transplants, implantation of medical devices and many other critical surgeries, where the risk of infection is high and any of these primary treatments could lead to an untreatable infection, leading to severe illness and significant loss of life (Kwon and Powderly, 2021).

A 2022 systematic analysis in the Lancet outlined the global burden of bacterial AMR in 2019 (Murray *et al.*, 2022). The authors estimated that 4.95 million deaths globally were associated with bacterial AMR in 2019 and that within these 4.95 million deaths there were an estimated 1.27 million deaths directly attributable to bacterial AMR. These data, analysing the global mortality caused by AMR five years after the first O'Neill publication, paint a bleak picture of our efforts to curb the spread of AMR in recent years prompting health experts, including those at the Centers for Disease Control and Prevention and the World Health Organization (WHO), to reiterate their warning that if we take life-saving antibiotics for granted and continue to overuse or misuse them, they will increasingly fail when we need them most.

1.5 How Antimicrobial Resistance Spreads Throughout Populations

Given the severe warnings from health experts and the alarming rate of AMR attributable deaths estimated to have increased from approximately 700,000 around 2014 to 1.27 million in 2019, it is important to understand how AMR spreads.

Antibiotics, used by both people and animals, enter the recipient's body where they will inhibit and/or kill bacterial populations. However, for some bacterial populations this exposure to the antimicrobial drug promotes the evolution of antibiotic resistance within the host. When people take antibiotics, they are often still well enough to move around, which can result in the spread of resistant pathogens to other people, animals or the environment. If the person administered with antibiotics is severely ill, they will probably be in a healthcare setting; in this case the resistant pathogens can spread throughout that healthcare setting, via transfer on surfaces, healthcare workers and other patients or visitors. The spread of resistant pathogens in hospitals is of particular concern due to the vulnerability of the patients (Avershina *et al.*, 2021). These resistant pathogens can also enter the general population as people carrying them leave the hospital.

Where animals are administered antibiotics, resistant bacterial species will evolve within the animal. In these cases, resistant bacteria can spread to humans, animals and within the environment directly by contact and indirectly via the food chain, water, air and manured soils (Marshall and Levy, 2011).

Exposure to antibiotics that drives resistance does not always drive resistance only for the antibiotic class used. In some cases, this exposure to a single antibiotic may drive the spread of resistance for multiple classes of antibiotics concurrently owing to the antibiotic-resistance genes being present on resistance plasmids. Resistance plasmids are small circular pieces of DNA that carry resistance to multiple different types of antibiotics, and they can be passed directly from one bacterial cell to another in proximity, for example within a biofilm. Exposure of a bacterial population to one antibiotic, where the resistance is found on a resistance plasmid, will select for the retention, and spread, of that plasmid, which therefore selects for the retention and spread of resistance genes to all antibiotics coded on that plasmid. It is therefore crucial to understand where and how resistance originates, to understand the spread of resistance within a given population or environment and to ensure that strategies employed to reduce the spread of that resistance are fit for purpose (Kruse and Sorum, 1994).

1.6 The Role of Biofilms in the Spread of Antimicrobial Resistance

There are many mechanisms for the spread of AMR; one of the most common means by which resistance spreads within a population is within biofilms. Biofilms are the dominant lifestyle for most microbes in the environment; the US National Institutes of Health state that amongst all microbial infections, around 80% are linked to biofilm formation. Biofilms form when free-swimming microbes attach to a surface, interface or other microbial cells and embed themselves in a self-producing matrix of polysaccharides, proteins and DNA, forming a community with altered phenotype (O'Toole *et al.*, 2000). As biofilms mature, they form complex structures and are organized in a manner that benefits the microbial community present. These structures allow the passage of water and nutrients to the cells embedded within the matrix and allow the formation of different microenvironments within the biofilm. This creates sub-populations of microbes within the same biofilm, with some entering a dormant state making them highly resistant to antimicrobials because many drugs utilize the metabolic processes of the bacteria to cause inhibitory effects. Resistance to antibiotics in biofilms is also mediated by the matrix due to reduced penetration of many drugs through the extracellular polymeric substances of the biofilm. These physical structures and properties of biofilms can make the microbes inside them up to 1000-fold more resistant to antimicrobials than their free-living counterparts (reviewed in Davies, 2003).

Biofilms are a major driver of the spread of antibiotic resistance due to the proximity of the microbes providing an excellent environment for the spread of antibiotic-resistance genes within the population. There are several mechanisms for the spread of AMR within a biofilm, including the direct passage of plasmids from one bacterial cell to another via conjugation, the spread of resistance genes by viruses and virus-like particles, shedding and envelopment of outer membrane vesicles and the uptake of raw DNA present in the biofilm matrix due to lysis of bacterial cells within the biofilm (Roberts *et al.*, 2014). When microbes leave the biofilm, either via natural dissemination mechanisms or mechanical disruption, they take the AMR genes that have been acquired with them into a new environment. This is a significant mechanism for spreading resistance between animals, people and the environment. If antibiotic treatments are also encountered within these environments, the spread of the antibiotic resistance can be further promoted.

Having established that biofilms are a significant mechanism for the spread of antibiotic resistance, it is important to understand the consequences of biofilm formation in the real world. In the healthcare sector alone, biofilm formation costs the UK over US$7 billion per year, with the global cost being closer to US$387 billion (Cámara *et al.*, 2022). Here, microbial attachment that leads to biofilm formation on medical devices such as catheters, replacement valves, artificial joints and central lines has significant impacts on quality of life, length of hospital stay and mortality. With the escalating prevalence of antimicrobial-resistant microbes inhabiting these biofilms, it is clear to see how the severity of infections arising from biofilm formation will increase dramatically.

The issue of biofilms is not limited to the healthcare sector, biofilms can form on almost any surface, which has huge impact on industries including food production, oil and gas production, clean water supplies and transportation (Cámara *et al.*, 2022). Biofilm contamination is a major global challenge, costing US$4 trillion per annum, exerting economic, social and environmental impacts.

With the estimation that around 80% of infections are caused by biofilm formation, the escalating rise in antibiotic resistance and the contribution of biofilms towards the spread of that resistance, clearly the consideration of biofilms should have a greater role in antibiotic stewardship programmes in future.

1.7 An Overview of the Current State of Antibiotic Usage in Food-producing Animals

All use of antibiotics contributes to the spread of antibiotic resistance, including use in human medicine and in raising animals. As a result, reductions in the use of antibiotics in human medicine alone is insufficient to tackle the spread of AMR, especially given that a significant percentage of antibiotics used around the world are sold for raising livestock. Unsurprisingly the amount of antibiotics consumed by animals varies considerably between different animals and within different countries. In the UK, data from 2017 showed that approximately two thirds of antibiotics sold were for human medicinal use, whilst approximately one third were used for animal care, including farm and companion animals (Veterinary Medicines Directorate, 2019). However, when global usage is explored the figures change dramatically. As recently as 2018, antibiotic sales for use in animals represented approximately 65% of all antibiotic sales within the USA (Wallinga and Kar, 2020) and further estimates placed the USA as the third highest veterinary consumer of antimicrobials in 2017 (~6000 tonnes), behind Brazil with just under 7400 tonnes and China with almost 42,000 tonnes of active antimicrobials (Tiseo *et al.*, 2020). However, many of these figures are difficult to determine accurately; for example, data from two independent sources in China stated antibiotic usage in 2017 as 41,967 tonnes with another estimating 78,200 tonnes was used. Disparate estimates such as these, and the fact that many countries do not currently record antibiotic usage, make understanding the true scale of antibiotic use globally extremely challenging.

Within the UK, the sale of veterinary antibiotics for use in food-producing animals decreased by approximately 53% between 2014 and 2018 and has remained largely stable since, reducing from 447 tonnes of active antibiotic ingredient sold for animal use in 2014 to 226 tonnes in 2020 (Veterinary Medicines Directorate, 2021). This reflects legislative requirements and a commitment from large parts of the industry within the UK to reduce antibiotic use in food-producing animals.

The overall consumption of antibiotics is an important factor in reducing the spread of resistance; however, it is also vital to monitor the usage rates of individual antibiotics, particularly those of human clinical significance. In 2016, the WHO revised their list of highest priority antibiotics that are critically important in human medicine for treating serious infections (WHO, 2016). This list includes quinolones, third generation and higher cephalosporins, macrolides, ketolides, glycopeptides and polymyxins. The 2021 UK VARSS Report showed that antibiotic use in food-producing animals was reduced in every class of antibiotic surveyed between 2014 and 2020, but significantly the sales of the highest priority critically important antibiotics – fluoroquinolones, third- and fourth-generation cephalosporins and colistin – were reduced by 69%, 80% and 62%, respectively, during this time (Veterinary Medicines Directorate, 2021).

There are significant differences observed in the 2021 VARSS Report between different animal industries. Whilst there is a general downward trend in antibiotic usage in animals, with the longer-term change in antibiotic sales being down in most areas, there is considerable variation between usage in different species. Usage in pigs reduced overall by 62% to 83.1 tonnes of active ingredient sold in 2020 and in turkeys, broilers and ducks the long-term reductions were 88%, 67% and 83% respectively, with a combined reduction to 21 tonnes of active ingredient used in 2020 (Veterinary Medicines Directorate, 2021). The only animal species that continues to show a long-term rise in the use of antibiotics according to the 2021 VARSS Report is salmon, which showed an 82% increase in sales up to 5.6 tonnes of active ingredient in 2020. However, for the WHO-classified highest-priority critically important antibiotics, there is an overall long-term trend of reduction in the use of these antibiotics for all species.

The data within the 2021 VARSS Report show that significant progress has already been made towards reducing antibiotic sales for food-producing animals (Veterinary Medicines Directorate, 2021). However, more must be done across all sectors to ensure a continued reduction in antibiotic usage, to preserve the activity of these life-saving drugs for future years.

1.8 Antimicrobial Stewardship; Current State, Barriers and Finding a Way Forward

The measures taken to prevent the spread of AMR are collectively known as antimicrobial stewardship. In reference to human medicine and according to the Centers for Disease Control and Prevention, antimicrobial stewardship is 'a system of informatics, data collection, personnel and policy/procedures that promote the optimal selection, dosing and duration of therapy for antimicrobial agents throughout the course of their use' to ensure responsible use for individual patient care and the longevity of antibiotic effectiveness (Centers for Disease Control and Prevention, 2021). The core elements of running this type of antibiotic stewardship programme include a commitment from leadership to dedicating the necessary human, financial and information technology resources to the programme, as well as the appointment of a single leader responsible for programme outcomes to ensure accountability. A single pharmacist leader should also be appointed responsible for working to improve antibiotic use. The expectation is that at least one recommended action will be implemented in the programme, such as systemic evaluation of ongoing treatment need after a set period of initial treatment to ensure that best practices are being followed. It is also vital that antibiotic prescribing and resistance patterns are monitored locally at the healthcare setting level as well as nationally and internationally. Regular reporting data on antibiotic use and resistance should be provided to doctors, nurses and relevant staff within healthcare settings and education of clinicians and healthcare workers about resistance, best practices and optimal prescribing is also key to programme success (Redfield and Khabbaz, 2019).

These programmes of antibiotic stewardship are proving somewhat effective in countries such as the USA and UK and, although they are not without their problems, progress is being made. Between 2016 and 2019 there was a 6.6% reduction in antibiotic consumption in the UK, with a further decrease of 10.9% between 2019 and 2020 (UK Health Security Agency, 2021). In countries such as the UK, there are clear mechanisms and infrastructures in place to protect the efficacy of antibiotics, including the requirement of a prescription to obtain antibiotics in most circumstances. However, many other countries have significant complications and barriers to successfully implementing antimicrobial stewardship programmes in the manner described above (Redfield and Khabbaz, 2018). Prescribing practices vary dramatically around the world; in many countries, antibiotics can be purchased by the end user without prescription and often without the end user having a clear understanding of whether and which antibiotics should be used, dosages, treatment duration, or the potential for the spread of AMR. It is clear to see why antibiotic stewardship programmes, such as those undertaken in the UK would be inappropriate for implementation in these countries. Here, different approaches should be taken based on the unique factors prevalent in each country, state and region.

Data from lower- and middle-income countries is limited, however recent global surveillance data show a significant AMR burden in these countries (Sharma *et al.*, 2022). Experience has shown that antimicrobial stewardship programmes following the core guidance above can be successfully implemented only when healthcare structures and procedures are available and in place. Therefore, a second essential set of core elements has been identified that can help countries build the necessary structures at the national, state and regional level to enable the implementation of successful healthcare facility antimicrobial stewardship programmes that consider the local context (Smith and Paulin, 2019).

Guidance on establishing antimicrobial stewardship interventions in lower-middle-income countries suggests that these programmes must be implemented via a stepwise approach, that builds on existing structures and reporting systems in place (if any exist). They should maximize teamwork and encourage personnel and clinical staff to participate by starting from small, simple and long-lasting actions. The implementation process requires a tailored approach to address the different factors that may influence antibiotic prescribing and use in a specific context, such as involving clinical staff in identifying local targets for improving antibiotic use, having a systematic approach to implementing antimicrobial stewardship interventions, reviewing progress over time and making changes when appropriate (Smith and Paulin, 2019). The implementation process should identify the behaviour of healthcare professionals and organizations to create evidence-based antimicrobial stewardship interventions because changing prescribing behaviour requires the consideration of factors that influence prescribing and using these factors to develop the action plan.

All antimicrobial stewardship strategies should include infection prevention and control measures to prevent patients and healthcare workers from being colonized with bacteria or acquiring infections. This action not only prevents healthcare-associated infections and deaths, but it also saves money, reduces the spread of AMR and supports high-quality health services (Taplitz *et al.*, 2017).

However, antibiotic stewardship programmes that consider human consumption alone would provide only part of the solution to the antibiotic resistance issue. It is vital that programmes are also implemented that focus on animal care, with many high-level strategies now taking a one-health approach to AMR that considers human, animal and environmental factors (Medicines Directorate, 2019). Antibiotic use in animals has been classified into three major categories by the WHO: therapeutic use, disease prevention and growth promotion. Therapeutic use refers to the use of antibiotics for treating animals with clinically diagnosed infectious diseases, for example the treatment of cows with mastitis. Disease prevention refers to the use of antibiotics in healthy animals considered to be at risk of infection; this includes the prevention of infectious diseases not yet diagnosed in groups of animals and controlling the spread of disease from one animal to another within a group. Growth promotion is the use of antibiotics, often at subtherapeutic concentrations, to increase the rate of weight gain and/or the efficiency of feed utilization in animals by other than purely nutritional means (WHO, 2017). Widespread antibiotic use drives the emergence of antibiotic-resistant microbes in food-producing animals, including organisms that can cause disease in humans, such as enterococci, *Escherichia coli*, campylobacters and salmonellae (Manyi-Loh *et al.*, 2018). The use of antibiotic growth promoters at low doses and for extended durations is particularly favourable for the selection of antibiotic-resistant microbes and exposure to a single agent can select for antibiotics in other classes via linkage of genes on mobile genetic elements such as plasmids.

Not surprisingly, many countries, such as those in Europe, have made substantial efforts to reduce antibiotic use in food-producing animals via the implementation of a range of stewardship interventions. These include the production of annual reports of antibiotic usage and resistance in humans and farm animals and antibiotic usage benchmarking to identify levels of antibiotic use in farms and by veterinarian prescribers (Patel *et al.*, 2020). Benchmarking of the larger pharmaceutical companies, who are best placed to enact real change, is also utilized with the co-operation of many large pharma groups. This includes the pharmaceutical companies setting antibiotic active ingredient discharge limits to third-party suppliers and developing strategies for the dissemination of these drugs in lower- and middle-income countries as well as finding alternate ways to incentivize sales personnel instead of by volume of antibiotic sales that promotes increased usage (Access to Medicine Foundation, 2021). Some countries have employed national targets for reducing antibiotic usage and the EU banned the use of antibiotics as growth promoters in 2006. New EU legislation that came into effect in January 2022 goes further, with a ban on the administration of antibiotics for prophylaxis and metaphylaxis (except in exceptional circumstances) in groups of animals; however, the UK government has, at the time of writing, yet to match this legislative commitment. A cornerstone of antibiotic stewardship in animal husbandry, as with human medicine, is the implementation of measures to reduce exposure of animals to pathogenic organisms through a variety of different mechanisms including good hygiene practices, development of animal housing design and build, group management and minimum moving and mixing and biosecurity measures (HM Government, 2019).

In 2017, the WHO, in collaboration with the Food and Agriculture Organization and the World Organization for Animal Health, released a global action plan for combating antibiotic resistance by reducing unnecessary antibiotic use in animals and humans. The WHO action plan is grounded in a one-health model, a 'collaborative, multisectoral, and transdisciplinary approach working at the local, regional, national, and global levels with the goal of achieving optimal health outcomes recognizing the interconnection between people, animals, plants, and their shared environment' (European Commission, 2017).

Finally, a transparent, objective assessment of trends in antibiotic usage and resistance patterns is crucial to understanding the impact of policy initiatives to enable the evaluation of strategies to determine the impact of each. The current evidence suggests that many of the current antibiotic stewardship measures have been associated with reductions in antibiotic resistance in food-producing animals (Patel *et al.*, 2020). For example, when Denmark banned the use of avoparcin, an antibiotic that shares chemical similarity to vancomycin, levels of vancomycin-resistant enterococci in both humans and livestock dropped within two years (Levy, 2014). Similarly, a systematic review of 81 studies found that interventions to decrease antibiotic use reduced the prevalence of antibiotic-resistant bacteria in animals by approximately 15% and multidrug-resistant bacteria by up to 32% (Tang *et al.*, 2017). Importantly for the food-producing animal industry, there is mounting evidence that antibiotic usage can be reduced without reducing farm productivity and profitability via the development and administration of new products such as organic acids, probiotics and vaccines that can reduce the need for antibiotic usage (Levy, 2014).

1.9 Conclusion

When considering antibiotic usage and stewardship programmes, it is vital to keep this stern warning from the Centers for Disease Control and Prevention and the WHO at the forefront of our minds: if we take life-saving antibiotics for granted and continue to overuse or misuse them, they will increasingly fail when we need them most.

References

Abraham, E.P. and Chain, E. (1988) An enzyme from bacteria able to destroy penicillin. *Reviews of Infectious Diseases* 10(4), 677–678.

Access to Medicine Foundation (2021) *Antimicrobial Resistance Benchmark Report 2021*.

Aldridge, S. and Sturchi, J.L. (1999) The Discovery and Development of Penicillin. Available at: https://www.acs.org/content/acs/en/education/whatischemistry/landmarks/flemingpenicillin.html (accessed 28 September 2022).

Avershina, E., Shapovalova, V. and Shipulin, G. (2021) Fighting antibiotic resistance in hospital-acquired infections: current state and emerging technologies in disease prevention, diagnostics and therapy. *Frontiers in Microbiology* 12, 707330.

Bengtsson-Palme, J., Jonsson, V. and Heß, S. (2021) What is the role of the environment in the emergence of novel antibiotic resistance genes? A modeling approach. *Environmental Science and Technology* 55, 15734–15743.

Blair, J.M.A., Webber, M.A., Baylay, A.J., Ogbolu, D.O. and Piddock, L.J.V. (2015) Molecular mechanisms of antibiotic resistance. *Nature Reviews Microbiology* 13, 42–51.

Bliven, K.A. and Maurelli, A.T. (2016) Evolution of bacterial pathogens within the human host. *Microbiology Spectrum* 4(1). doi: 10.1128/microbiolspec

Bottery, M.J., Pitchford, J.W. and Friman, V.P. (2021) Ecology and evolution of antimicrobial resistance in bacterial communities. *EFSA Journal* 15(4), 939–948.

Cámara, M., Green, W., MacPhee, C.E., Rakowska P.D., Raval, R. *et al.* (2022) Economic significance of biofilms: a multidisciplinary and cross-sectoral challenge. *Biofilms and Microbiomes* 8(1), 1–8.

Castanon, J.I.R. (2007) History of the use of antibiotic as growth promoters in European poultry feeds. *Poultry Science* 86(11), 2466–2471.

Centers for Disease Control and Prevention (2021) *Core Elements of Antibiotic Stewardship*. Available at: https://www.cdc.gov/antibiotic-use/core-elements/index.html (accessed 28 September 2022).

Davies, D. (2003) Understanding biofilm resistance to antibacterial agents. *Nature Reviews Drug Discovery* 2(2), 114–122.

EFSA (2021) Third joint inter-agency report on integrated analysis of consumption of antimicrobial agents and occurrence of antimicrobial resistance in bacteria from humans and food-producing animals in the EU/EEA. *EFSA Journal* 19(6), 6712.

European Commission (2017) A European One Health Action Plan against Antimicrobial Resistance (AMR). Available at: https://health.ec.europa.eu/system/files/2020-01/amr_2017_action-plan_0.pdf (accessed 28 September 2022).

Fleming, A. (1945) Sir Alexander Fleming. Nobel Lecture. Available at: https://www.nobelprize.org/prizes/medicine/1945/fleming/lecture/ (accessed 15 September 2022).

Gatermann, S., Das, S., Dubreuil, L., Giske, C.G., Kahlmeter, G. *et al.* (2022) Expected phenotypes and expert rules are important complements to antimicrobial susceptibility testing. *Clinical Microbiology and Infection* 28(6), 764–767.

HM Government (2019) Tackling antimicrobial resistance 2019 to 2024. Available at: https://www.gov.uk/government/publications/uk-5-year-action-plan-for-antimicrobial-resistance-2019-to-2024 (accessed 28 September 2022).

Hutchings, M., Truman, A. and Wilkinson, B. (2019) Antibiotics: past, present and future. *Current Opinion in Microbiology* 51, 72–80.

Karkman, A., Pärnänen, K. and Larsson, D.G.J. (2019) Fecal pollution can explain antibiotic resistance gene abundances in anthropogenically impacted environments. *Nature Communications* 10(1), 1–8.

Klug, D.M., Idiris, F.I.M., Blaskovich, M.A.T., von Delft, F., Dowson, C.G. *et al.* (2021) There is no market for new antibiotics: This allows an open approach to research and development. *Wellcome Open Research* 6(146), 1–14.

Kristiansson, E., Fick, J., Janzon, A., Grabic, R., Rutgerssen, C. *et al.* (2011) Pyrosequencing of antibiotic-contaminated river sediments reveals high levels of resistance and gene transfer elements. *PLoS ONE* 6(2), e17038.

Kruse, H. and Sorum, H. (1994) Transfer of multiple drug resistance plasmids between bacteria of diverse origins in natural microenvironments. *Applied and Environmental Microbiology* 60(11), 4015–4021.

Kwon, J.H. and Powderly, W.G. (2021) The post-antibiotic era is here. *Science* 373(6554), 471.

Ledger, E.V.K., Sabnis, A. and Edwards, A.M. (2022) Polymyxin and lipopeptide antibiotics: membrane-targeting drugs of last resort. *Microbiology* 168, 1136.

Levy, S. (2014) Reduced antibiotic use in livestock: How Denmark tackled resistance. *Environmental Health Perspectives* 122(6), 160–165.

Li, D., Yang, M., Hu, J., Ren, L., Zhang, Y. *et al.* (2008) Determination and fate of oxytetracycline and related compounds in oxytetracycline production wastewater and receiving river. *Environmental Toxicology and Chemistry* 27(1), 80.

Lowy, F.D. (2003) Antimicrobial resistance: the example of *Staphylococcus aureus*. *Journal of Clinical Investigation* 111(9), 1265–1273.

Manyi-Loh, C., Mamphweli, S., Meyer, E. and Okoh, A. (2018) Antibiotic use in agriculture and its consequential resistance in environmental sources: potential public health implications. *Molecules* 23(795), 1–48.

Marshall, B.M. and Levy, S B. (2011) Food animals and antimicrobials: Impacts on human health. *Clinical Microbiology Reviews* 24(4), 718–733.

Matange, N., Hegde, S. and Bodkhe, S. (2019) Adaptation through lifestyle switching sculpts the fitness landscape of evolving populations: implications for the selection of drug-resistant bacteria at low drug pressures. *Genetics* 211, 1029–1044.

McKenna, M. (2013) Imagining the post-antibiotics future. Available at: https://thefern.org/2013/11/imagining-the-post-antibiotics-future/ (accessed 28 September 2022).

Melnyk, A.H., Wong, A. and Kassen, R. (2014) The fitness costs of antibiotic resistance mutations. *Evolutionary Applications* 8, 273–283.

Murray, C.J., Ikuta, K.S., Sharara, F., Swetschinski, L., Aguilar, G.R. *et al.* (2022) Global burden of bacterial antimicrobial resistance in 2019: a systematic analysis. *The Lancet* 399(10325), 629–655.

O'Neill, J. (2014) Antimicrobial Resistance: Tackling a crisis for the health and wealth of nations. Available at: https://amr-review.org/sites/default/files/AMR%20Review%20Paper%20-%20Tackling%20a%20crisis%20for%20the%20health%20and%20wealth%20of%20nations_1.pdf (accessed 28 September 2022).

O'Toole, G., Kaplan, H.B. and Kolter, R. (2000) Biofilm formation as microbial development. *Annual Review of Microbiology* 54, 49–79.

Partridge, S.R., Kwong, S.M., Firth, N. and Jenson, S.O. (2018) Mobile genetic elements associated with antimicrobial resistance. *Clinical Microbiology Reviews* 31(4), e00088–17.

Patel, S. (2016) Drivers of bacterial genomes plasticity and roles they play in pathogen virulence, persistence and drug resistance. *Infection, Genetics and Evolution* 45, 151–164.

Patel, S.J., Wellington, M., Shah, R.M. and Ferreira, M.J. (2020) Antibiotic stewardship in food-producing animals: challenges, progress, and opportunities. *Clinical Therapeutics* 42(9), 1649–1658.

Rammelkamp, C.H. and Maxon, T. (1942) Resistance of *Staphylococcus aureus* to the action of penicillin. *Experimental Biology and Medicine* 51(3), 386–389.

Redfield, R.R. and Khabbaz, R. (2018) The Core Elements of Human Antibiotic Stewardship Programs in Resource-Limited Settings: National and Hospital Levels. *Atlanta.* Available at: https://www.cdc.gov/antibiotic-use/healthcare/implementation.html (accessed 23 March 2023).

Redfield, R.R. and Khabbaz, R. (2019) Core Elements of Hospital Antibiotic Stewardship Programs. Atlanta. Available at: https://www.cdc.gov/antibiotic-use/core-elements/hospital.html (accessed 23 March 2023).

Roberts, A.P., Kreth, J. and Jakubovics, N.S. (2014) The impact of horizontal gene transfer on the adaptive ability of the human oral microbiome. *Frontiers in Cellular and Infection Microbiology* 4, 124.

Sharma, A., Singh, A., Dar, M.A., Kaur, R.J., Charan, J. *et al.* (2022) Menace of antimicrobial resistance in LMICs: Current surveillance practices and control measures to tackle hostility. *Journal of Infection and Public Health* 15(2), 172–181.

Smith, I. and Paulin, S. (2019) Antimicrobial stewardship programmes in health-care facilities in low- middle-income countries. A WHO practical toolkit. Geneva. Available at: https://apps.who.int/iris/handle/10665/329404 (accessed 28 September 2022).

Tang, K.L., Caffrey, N.P., Nobrega, D.B., Cork, S.C., Ronksley, P.E. *et al.* (2017) Restricting the use of antibiotics in food-producing animals and its associations with antibiotic resistance in food-producing animals and human beings: a systematic review and meta-analysis. *The Lancet Planetary Health* 1(8), e316–e327.

Taplitz, R.A., Ritter, M.L. and Torriani, F.J. (2017) Chapter 6: Infection Prevention and Control, and Antimicrobial Stewardship. In: *Elsevier Public Health Emergency Collection: Infectious Diseases.* Elsevier, Amsterdam, pp. 54–61.e1.

Thomas, D. and Wessel, C. (2022) The State of Innovation in Antibacterial Therapeutics. Available at: https://www.bio.org/sites/default/files/2022-02/The-State-of-Innovation-in-Antibacterial-Therapeutics.pdf (accessed 28 September 2022).

Tiseo, K., Huber, L., Gilbert, M., Robinson, T.P., Van Boeckal, T.P. *et al.* (2020) Global trends in antimicrobial use in food animals from 2017 to 2030. *Antibiotics* 9(12), 1–14.

UK Health Security Agency (2021) English surveillance programme for antimicrobial utilisation and resistance (ESPAUR) Report 2020 to 2021. Available at: https://assets.publishing.service.gov.uk/government/uploads/system/uploads/attachment_data/file/1069632/espaur-report-2020-to-2021 16-Nov-FINAL-v2.pdf (accessed 28 September 2022).

Veterinary Medicines Directorate (2019) UK One Health Report – Joint report on antibiotic use and antibiotic resistance, 2013–2017. Available at: https://assets.publishing.service.gov.uk/government/uploads/system/uploads/attachment_data/file/921039/Ted_Final_version__1318703-v45-One_Health_Report_2019_FINAL-accessible.pdf (accessed 23 March 2023).

Veterinary Medicines Directorate (2021) UK Veterinary Antibiotic Resistance and Sales Surveillance Report. Available at: https://www.gov.uk/government/publications/veterinary-antimicrobial-resistance-and-sales-surveillance-2021 (accessed 23 March 2023).

Wallinga, D. and Kar, A. (2020) New Data: Animal vs. Human Antibiotic Use Remains Lopsided. Available at: https://www.nrdc.org/experts/david-wallinga-md/most-human-antibiotics-still-going-us-meat-production (accessed 28 September 2022).

WHO (2016) Critically Important Antimicrobials for Human Medicine 5th Revision 2016 Ranking of medically important antimicrobials for risk management of antimicrobial resistance due to non-human use. Available at: https://www.who.int/publications/i/item/9789241515528 (accessed 28 September 2022).

WHO (2017) WHO guidelines on use of medically important antimicrobials in food-producing animals. Available at: http://apps.who.int/iris/bitstream/handle/10665/258970/9789241550130-eng.pdf (accessed 28 September 2022).

Yu, D., Zheng, Y. and Yang, Y. (2020) Is there emergence of β-Lactam antibiotic-resistant *Streptococcus pyogenes* in China? *Infection and Drug Resistance* 13, 2323–2327.

CHAPTER 2
Moving Beyond Antibiotic Usage to Support the Health and Productivity of Commercially Reared Poultry

Roberto La Ragione[1] and Emily Burton[2]*
[1]University of Surrey, UK; [2]Nottingham Trent University, UK

Abstract

The global demand for poultry meat and eggs continues to grow and the industry is expected to expand at a rapid pace. Infectious diseases are responsible for significant economic losses in the poultry industry and have implications with regards to welfare and economics. Furthermore, a number of poultry pathogens are zoonotic and thus pose a public health risk. Traditionally, infectious diseases of poultry have been controlled through the use of vaccination, biosecurity measures and antibiotics. However, with the increased awareness of the emergence of antimicrobial resistance (AMR) and the ban of antibiotic growth promoters in many countries, alternative control strategies are urgently required. Alternatives to antibiotics may include the use of vaccines, novel antimicrobials, prebiotics and probiotics to modulate the gut microflora. Thus, this chapter will focus on why alternatives are required and the development of novel alternatives to antimicrobials for use in the poultry industry.

Phytochemicals, including polyphenols, and metal complexes such as carbon monoxide-releasing molecules are novel compounds that alone or in combination have been shown to be effective against a vast array of human and animal bacterial pathogens. Probiotics are often members of the normal flora and *Lactobacillus*-based probiotics have been reported previously as protecting against infection with common enteric bacterial pathogens in livestock and poultry. Moreover, prebiotics such as galactooligosaccharide have been shown to increase the number of bacteria such as bifidobacteria and lactobacilli and/or their fermentation products in the intestine. The mechanisms by which

*Corresponding author: emily.burton@ntu.ac.uk

DOI: 10.1079/9781800622746.0002

these novel interventions exert their antimicrobial activity are multifactorial, but may include modulation of the intestinal flora, thus creating an inhospitable environment for pathogens. Collectively, novel interventions may be effective alternatives to antibiotics for the control of bacterial pathogens in poultry and thus reduce reliance on the use of antibiotics.

2.1 Historic Use of Antibiotics in Animal Feed

In response to continual pressure on global demand for animal protein, poultry meat production levels continue to grow year on year (OECD/FAO, 2021). While antibiotic consumption patterns in agriculture vary markedly across regions and countries (WHO, 2018), resistance to antibiotics is an inherent side effect associated with the inappropriate use of antibiotics (Williams-Nguyen *et al.*, 2015). Compounds with antibiotic activity to either inactivate microorganisms or inhibit their growth and replication may be employed in several different ways. Alongside antiseptics and disinfectants, antibiotics form part of the armoury of antimicrobials widely used to treat infections in humans and animals. However, beyond their use in the clinical treatment of infectious disease in people and animals ('therapeutic use'), antibiotics have been used to prevent or control common disease events ('prophylactic use') or at sub-therapeutic levels over long periods of time to improve daily weight gain and feed efficiency through alterations in digestion and disease suppression ('antibiotic growth promotion'). Antibiotics may also be used to treat a single animal with clinical disease or a group of animals (metaphylaxis). However, the definitions associated with these usage terms often vary by sector, governing body and geographic region and may be applied concurrently within a given livestock population (Landers *et al.*, 2012). For example, in 1999, 16% of all US dairy cows were reported to be receiving antibiotic treatment for clinical mastitis, while almost all dairy cows are given prophylactic treatment following each lactation as a preventive measure against mastitis (NRC, 1999). However, in recent years this has changed significantly with restrictions on antibiotic use and better awareness regarding the development and spread of resistance.

The lack of distinction between type of antibiotic usage has made the creation of policies and guidelines for the addition of antibiotics to the animal feed a global challenge for many years (Chattopadhyay, 2014). Concerns over the use of antibiotics as antibiotic growth promoters were noted in the UK House of Lords Swann Report (Swann, 1969) within 15 years of Jukes and Williams (1953) recognizing their potential usage to promote efficiency in poultry and livestock production. The use of antibiotics as growth promoters in animal feeds was permitted in the member states of the European Union (EU) for more than 50 years and still occurs in some countries.

However, concerns regarding the development of AMR and the transfer of antibiotic-resistance genes from animal to human microflora led to the withdrawal of approval for antibiotics as growth promoters in the EU from January 2006 (Castanon, 2006). Prior to this, Sweden was the first country to legislate against the use of all growth-promoting antibiotics in 1996, followed by more specific legislation in Denmark to ban use of avoparcin and virginiamycin in 1995 and 1998, respectively. At this time, elsewhere in the EU, scientific concerns were growing that bacteria harbouring antibiotic-resistant genes were being inadvertently selected in animal hosts through antibiotic treatment and might be transmitted to humans. Alongside consumer and political opinion, this concern led the EU to ban the use of avoparcin in 1997 and the four remaining antibiotics used for growth promotion in 1999. However, reports on the Swedish approach began to reveal the scale of negative impacts associated with the ban (Wierup, 2001); although these antibiotics were originally approved for use as growth promotors, their sudden withdrawal revealed their concurrent role in prophylaxis of infectious disease (Casewell *et al.*, 2003). The dramatic increase in infections led to a substantial increase in the use of therapeutic antibiotics for livestock and poultry in Europe (Larsen, 2002).

There is now consensus across a number of global regions that antibiotics should only be utilized for the control and treatment of infections in animals and not for the promotion of growth, increased performance and improved feed efficiency (FDA, 2012). Despite this consensus, the extensive beneficial effects obtained from antibiotic use still make antibiotic treatment an attractive option for the profitable and efficient production of livestock until the long-term threat to humankind outlined in the opening chapter is considered. Therefore, many global regions are creating more stringent legislation to restrict the routine use of antibiotics in agriculture; in 2022, Regulation (EU) 2019/6 on veterinary medicines and Regulation (EU) 2019/4 on Medicated Feed came into force to prohibit preventative group treatments.

2.2 What Do We Mean by Alternatives to Antibiotic Growth Promotors (AGPs)?

The means by which antibiotics exert their growth promoting effects are multifactorial (Castanon, 2007). In addition to inhibiting sub-clinical infections, antibiotics profoundly alter the intestinal environmental by removing the resident microbial population. This reduces the presence of growth-depressing microbial metabolites (Le Roy *et al.*, 2019) and enhanced uptake and use of nutrients by the host due to the reduced microbial use of nutrients (Anderson *et al.*, 2000). These traits cannot be mimicked by non-antibiotic compounds so alternative modes of action are required, of which the key trait is rapid development of a healthy gut microflora. Several other factors may also influence whether compounds are considered viable alternatives to antibiotic growth promoters (AGPs) including stabilization of digestion; stimulation and rapid maturation of the immune system; increased growth performance; reduced incidence of diarrhoea; improved feed conversion efficiency; lower mortality and ultimately higher profitability. There are a number of agents/compounds that exhibit many of these properties including probiotics, prebiotics, synbiotics, postbiotics, next-generation growth promoters, metals, phage therapy, natural plant extracts, vaccines, acidifiers, enzymes and faecal transplants (Betts *et al.*, 2018).

As we produce more animals for food to meet the continual pressure on global demand, levels of AMR associated with animal production will concomitantly rise unless alternative strategies are adopted to maintain gut health. Five key bacterial pathogens where alternative control strategies are urgently required are *Salmonella*, *Campylobacter*, *Escherichia coli*, *Clostridia* and *Brachyspira*. Many of these pathogens are zoonotic and thus can infect humans and also transfer AMR from animals to humans.

The global public health threat from AMR led the World Health Organization (WHO) to establish guidelines in 2018 to preserve the effectiveness of the highest priority, critically important antimicrobials (HPCIAs). Based on clinical importance in human medicine and the risk of resistance transfer from animals, the WHO listed five classes of HPCIAs: extended-spectrum cephalosporins (also known as 3rd, 4th and 5th generation cephalosporins); macrolides and ketolides; glycopeptides; polymyxins and quinolones (Aidara-Kane *et al.*, 2018). Although the WHO has recommended that drugs classified as HPCIAs for human medicine should not be used for the treatment of food-producing animals even where infectious disease is clinically diagnosed, compliance with the guidelines is geographically inconsistent and difficult to ascertain owing to poor reporting in some regions (WHO, 2017). An investigation into the use of HPCIAs to treat infectious disease in production and companion animals reported the most widely used HPCIAs in poultry production were enrofloxacin, colistin and macrolides, with considerable variation from country to country (Lhermie *et al.*, 2020).

Alongside breeding and feeding strategies to reduce antimicrobial use in combatting bacterial and parasitic infections in poultry gastrointestinal tracts, vaccines are also emerging as a front-line control tool (Chambers *et al.*, 2016). To be effective in the long term, however, vaccines need to be deployed alongside comprehensive control strategies such as improved biosecurity (Davies and Wales, 2019). Navigating the options along the path of vaccine development can also be highly challenging due to the need to ensure the requirements for acceptable cost, practicality, efficacy and legislative compliance are met.

The development of commercial strains of poultry not only for productivity but also for disease resistance is now a key focus of the poultry industry (Bovenhuis *et al.*, 2022; Long *et al.*, 2019). In addition to the disease resistance emerging from global breeding programmes, feed additives offer a major route to improve gut health in poultry. However, a barrier to the broad uptake of some of the compounds described in this chapter is that each additive is not supplied with the structured body of evidence that accompanies licensed drugs (Ritter, 2008). Therefore, rigorously designed and conducted studies that examine not only the efficacy, but also the mode of action behind candidate feed additives in combatting bacterial infections of the gastrointestinal tract are essential to build understanding and sustained confidence in their practical application.

2.3 Benefits of Modulating the Avian Gut Microflora

The potential for feed additives to invoke an immunostimulatory response in the host bird, and at the same time help to ensure that flocks are only positively impacted, is a key area of nutritional interest. In particular, the relationship between the inclusion of additives in feed and their interaction with the intestinal microbe population is now considered a main route to improved conversion ratios and reduced incidence of disease and, consequently, economic benefit to the farm. During the AGP era, an intestinal microbial population was simply viewed as a drain on incoming nutrient resources, with no role for the intestinal microflora in maintaining host health. However, without the intestinal presence of antibiotic substances as killing agents, the bacteria that are present in a healthy poultry intestine are now viewed as a route to reduce the presence and virility of pathogenic bacteria. Later chapters in this book explore not only the routes by which the intestinal microflora improve the protective barrier functions of gastrointestinal tract, but also how the intestinal microflora support the host in extracting further nutritional benefit from the diet.

The intestinal microflora is not only a major utilizer of host-consumed nutrients; they can also contribute vitamins and biofactors in sufficient volume to support host health. While this field has been explored in mammals, the quantitative impact of the microflora on nutrient supply and demand is generally not well understood in poultry beyond some of the specific areas described in later chapters. In particular, the potential for the microflora to synthesize and supply vitamins to the host has been little studied in modern commercially used strains of laying hens, broiler chickens and turkeys. However, a recent review has considered the theoretical contribution of the microflora to nutrient supply and posed values for B-vitamin supply in laying hens, broilers and turkeys (Korver, 2022). This was achieved by applying values for theoretical B-vitamin production by the human intestinal microbiome (Magnúsdóttir *et al.*, 2015) and creating corresponding estimates of microbial and dietary vitamin supply in poultry. The author noted, however, that their assumptions were based on a mature, stable microflora that is rarely seen in meat poultry less than 25 days post-hatch and also that the retrograde peristalsis in the avian digestive tract (Brummermann and Braun, 1995) makes it likely that intestinal absorption of microbially synthesized vitamins from the caeca may be more efficient in poultry than in pigs or humans. The diet of the host has also been reported as highly influential: vitamin synthesis pathways for pyridoxine, riboflavin, vitamin K and vitamin E were all enriched by a dietary *Macleaya cordata* extract (Huang *et al.*, 2018).

Therefore, the modulation of the microflora is a key route to the healthy gut associated with not only optimal diet digestion, but also stimulation and rapid maturation of the immune system in order to protect the host against infectious disease. Through these two mechanisms, maintenance of optimal gut health leads to increased growth performance, reduced incidence of disease such as diarrhoea, improved efficiency, lower mortality rates and, ultimately, higher profitability and lowered use of global resources to produce poultry meat.

2.4 Influence of Viral Infections on the Microflora

The interplay between pathogens and host microflora plays an important role in health and disease in many vertebrates (Yitbarek *et al.*, 2018). Understanding changes in the host microflora resulting from viral infections is particularly of interest, as it could provide information for developing new methods for infectious disease prevention and treatment. Studies in the chicken model have shown compositional changes in gut microflora and differences in some of immune gene expression levels following H9N2 avian influenza virus (AIV), Newcastle disease virus (NDV) or infectious bronchitis virus (IBV) infection (Cui *et al.*, 2018; Xu *et al.*, 2019). However, there are many environmental factors, such as age, breed, diet, housing, hygiene and temperature, that can also affect the chicken microflora and thus might change the interactions between the host and viruses during the time of infection. To begin exploring the effects of viral infection on the poultry intestinal microflora, a study was undertaken using two different chicken breeds, Rhode Island Red and VALO leghorns, reared in different facilities, to identify commonality and differences in the host microflora changes during and following infection of these two chicken lines with H9N2 AIV (Chrzastek *et al.*, 2021). The authors reported that taxonomic analysis showed significant changes in microbial diversity and abundance in the healthy colon of chickens following low pathogenic avian influenza virus infection regardless of the chicken line infected.

2.5 Influence of Antibiotic Treatment on Gut Health and Metabolism

Although for many years now the European poultry sector has strived to reduce the administration of antibiotics in poultry production, there are disease situations where antibiotic treatment is unavoidable. An unfortunate consequence of antibiotic treatment is the reduction in gut microflora diversity and destabilization of the gut microflora, which in turn disrupts the symbiotic relationship existing between the host and its gut microbes (Spor *et al.*, 2011) and leads to metabolic weakening of the host (Antharam *et al.*, 2013). The response of the gut microflora to Tiamulin™, an antibiotic of the pleuromutilin family, following avian intestinal spirochaetosis (AIS) has been studied to determine the long-term effect of antibiotic treatment on the intestinal microflora of laying hens.

The study employed high-resolution 1H nuclear magnetic resonance (NMR) spectroscopy to explore the metabolic effect of both *Brachyspira pilosicoli* and Tiamulin on the host (chicken) systemic metabolism, while concurrently investigating the effects of both pathogen and drug treatment on the gut microflora composition using 16S rRNA next generation sequencing (NGS). The performance deterioration observed in untreated birds was partly explained by modification of systemic host energy metabolism as glycerol became adopted as the main glucose precursor. However, antibiotic treatment itself triggered an increased very-low-density lipoprotein (VLDL)/ high-density lipoprotein (HDL) ratio in plasma that may also contribute to reducing weight loss observed in treated birds. As expected, exposure to both the intestinal pathogen and antibiotic resulted in a profound shift in the microbial profile of laying hens likely to trigger dysbacteriosis (Le Roy *et al.*, 2019).

2.6 Understanding Microbial Diversity in the Poultry Gut through Next-generation Sequencing

The well-documented planetary dominance of the chicken in terms of biomass (Bennett *et al.*, 2018) has made the chicken gut microbiome one of the most abundant microbial communities on the planet. The chicken microbiome is dominated by bacteria but contains a rich variety of microorganisms with a taxonomic community shown to vary with age, breed, setting and disease status (Rychlik, 2020). Documenting this diversity in order to identify new opportunities for the improvement of poultry health and production has been a consistent research aim for some years (Shang *et al.*, 2018). However, approaches such as 16S rRNA gene sequencing, which rely largely on analyses of molecular barcodes, are unable to reveal details on the genome sequences, population structures or functional repertoires of microbial species (Hillmann *et al.*, 2018). In order to explore taxonomic and functional diversity in the complex microbial communities found in poultry, a metagenomic assembly approach has been recently developed. Gilroy *et al.* (2021) exploited two complementary approaches – metagenomics and culture – to create an extensive catalogue of genes, genomes and isolates from this important ecosystem. Data from Shotgun metagenomic sequencing of DNA extracted from relevant samples, followed by bioinformatics-based community profiling and analysis (Glendinning *et al.*, 2020; Sergeant *et al.*, 2014) were combined with large-scale isolation of microorganisms in pure culture with whole-genome sequencing and phylogenomic analysis (Medvecky *et al.*, 2018) and used to generate phylogenetic profiles. These profiles were then used to document known and unknown diversity to create a high-quality reference collection of microbial genes and genomes from the chicken gut, including hundreds of new candidate species that were allocated well-formed Latin binomials and uploaded into public genome repositories. This type of approach substantially improves the coverage of the chicken gut microbiome in the public databases and will make it possible to profile sequences from the chicken gut much more rapidly, easily and comprehensively. Once fully established, databases such as these provide a valuable resource for assessing the effects of pre and probiotic candidate feed additives on intestinal modulation in poultry.

Approaches combining *in silico* phylotyping and core genome single nucleotide polymorphism analysis have also been employed to advance specific research challenges such as reducing the impact of avian pathogenic *E. coli* (APEC) on laying poultry flocks. The establishment of *E. coli* as both a model organism and potential pathogen has diverted attention away from commensal strains and other species in the *Escherichia* genus. The vast array of information on some members of the *Escherichia* genus and the lack of whole-genome sequencing studies including all members of the genus skews our understanding of the genus, making it difficult to advance our understanding of the population structure and ecology of *Escherichia* (Lagerstrom and Hadly, 2021). The investigation of faecal samples by Thomson *et al.* (2022), taken from a small flock of healthy layer chickens, revealed remarkable genomic diversity among *Escherichia* isolates: representatives of almost the entire known phylogeny of *Escherichia* were recovered, including species that have only recently been recognized (previously known as cryptic clades) and for which chicken-associated isolates have not been widely reported. These isolates had low carriage rates of AMR genes and virulence factors, suggesting that similar isolates might often be overlooked in studies that focus on these traits. The findings highlight the surprising diversity of *Escherichia* harboured by even an individual chicken and emphasize the need to broaden the focus of research to encapsulate the full variety of species.

In addition to use of modern sequencing approaches to characterize the intestinal microbiome composition of broilers as a tool for assessing the effects of gut-modulating feed additives, they are currently playing a direct role in efforts to reducing the levels of food-borne disease affecting human health. Pathogenic *Campylobacter* species *Campylobacter jejuni* and *Campylobacter coli*, derived from handling or ingesting undercooked chicken meat, are common causative agents of bacterial gastroenteritis in humans (Nichols *et al.*, 2012). Alongside improved biosecurity efforts to reduce the occurrence of *Campylobacter* at poultry farms, reducing the prevalence of *Campylobacter* contamination in chicken meat could be achieved by decreasing the on-farm prevalence either by reducing the number of infected flocks and/or reducing levels within individual birds.

Once *Campylobacter* is present within a poultry shed, there is rapid transmission within the flock, but the ability of a given strain to colonize the broiler gut depends on the strain characteristics, the immune status and health of the bird and the composition of the existing the gut microbiome (Qu *et al.*, 2008). Detailed knowledge of the microbiome before and during *Campylobacter* colonization is likely to inform the development of interventions such and pre and probiotic feed additives aimed at reducing the prevalence of *Campylobacter* in broilers. The outcomes of a study employing advanced sequencing techniques by Sakadiris *et al.* (2018) suggest that there is strong evidence that high counts of *Campylobacter* in broiler caeca are interrelated with the microbial community structure and particularly associated with an increase in the presence of Enterobacteriaceae, but it remains unclear whether the presence of *Campylobacter* favours the growth of Enterobacteriaceae or vice versa. It is possible that a general dysbacteriosis leads to colonisation by both *Campylobacter* and Enterobacteriaceae. Furthermore, it was found that a decrease in *Lactobacillus* abundance in chicken caeca could be linked to high *Campylobacter* loads. Therefore, although *Campylobacter* are commonly regarded as a benign commensal bacterium in poultry (USDA, 2022), their association with increased prevalence of dysbacteriosis suggests levels of *Campylobacter* contamination of meat may reduce with increased efforts to optimize the intestinal microbiome for maximum poultry health and productivity.

2.7 Use of *In Vitro* Gut Models for Assessing Candidate Gut Health Enhancing Products

In vitro gut models provide a powerful screening tool that can be used to assess and refine interventions that mitigate the spread of antibiotic resistance in the gut before undertaking animal studies. Spreading of AMR via plasmid exchange is a transfer route of particular concern as it enables unrelated bacteria to acquire resistance. Card *et al.* (2017) used a chemostat-based system to model the effects of bacterial infection, antibiotic administration and plasmid transfer in chicken gut microbiota and demonstrated the transfer of a multidrug-resistant plasmid from the zoonotic pathogen *Salmonella* to commensal *E. coli* occurs at a high rate, even in the absence of antibiotic administration.

To analyse accurately the impact of external factors on both bacterial communities within the gastrointestinal tract and their fermentation products, *in vitro* models are required. However, many *in vitro* fermentation models are limited in their usefulness as they fail to represent accurately the dynamic flow of digesta through the gastrointestinal tract (Murray *et al.*, 2014). An advantage of *in vitro* models over post-mortem or fistulated *in vivo* sampling approaches, however, is that closed systems support the quantification of short-chain fatty-acid products from bacterial fermentation. In particular, research into the use of prebiotics such as mannan oligosaccharides (Lea *et al.*, 2013), or more recently, xylooligosaccharides (De Maesschalck *et al.*, 2015), to stimulate the overall butyrate-producing capacity of intestinal microflora benefit from *in vitro* models. The development of *in vitro* fermentation systems that incorporate the modelling of digesta flow would retain the vital quantification of fermentation products associated with traditional *in vitro* bacterial fermentation models while increasing their relevance to the *in vivo* use of feed additives in poultry.

Leng *et al.* (2019) developed a three-stage fermentation model designed to simulate bacterial communities in the equine large intestine. The model was used for concordance with *in vivo* gut bacterial populations, repeatability and ability to capture individual variation in the bacterial community profile. A concordance study comprising three equine gut models showed that the faecal bacteria introduced into the gut model were able to metabolize nutrients within the culture medium to produce volatile fatty acids but the abundance of bacterial populations were substantially altered immediately after inoculation and, while the majority of populations rapidly stabilized to levels reflecting those found *in vivo*, not all the populations present in gut content samples were restored in the model on reaching steady state. Subsequence studies have indicated that the equine gut model may have some capacity to support exploration of the interindividual variation seen in the gut microflora of equine populations, which may also have relevance to poultry production. The repeatability study provided preliminary evidence that models inoculated with the same faecal matter will produce microbial communities that are both structurally and functionally similar. The findings of this study suggest a three-stage fermentation model is extremely useful in exploring the impact of diet and other external factors on the intestinal microflora of poultry as the system is now capable of maintaining the ratio of organisms found in live host microbiomes. Currently, the model system is being applied to explore what happens to antibiotic transfer and particularly how plasmids that confer resistance transfer from one organism to another.

2.8 Emerging Feed-based Alternatives to Therapeutic and Prophylactic Use of Antibiotics in Poultry

Alongside our increasing understanding of the impact of removing from the feed antibiotic substances contributing a blended AGP and prophylactic role, the poultry sector is beginning to accept that there will be no single replacement for the role played by antibiotics in feed. Previous and subsequent chapters in this book outline the type of multifactorial strategies required to achieve levels of production efficiency and diseases incidence similar to those recorded in the AGP usage era. The remainder of this chapter explores the role of some established and more emerging pre and probiotic approaches to improving poultry health and productivity in the post-AGP era.

2.9 Characterization of Probiotic Bacteria to Combat Infectious Diseases in Poultry

Almost as intensive meat poultry became established after World War II, researchers observed that *Salmonella* infection in birds was suppressed with increasing age (Milner and Shaffer, 1952). The theory of competitive exclusion of pathogenic bacteria through the establishment of a benign residential flora soon followed (Nurmi and Rantala, 1973). Mechanisms employed during competitive exclusion by residential microbes include the secretion of antimicrobial compounds, competition for essential nutrients and host cell-binding receptors and immunomodulation (Vanderpool *et al.*, 2008). Specifically, lactobacilli can exclude *Campylobacter* (Zhang *et al.*, 2007), *Clostridium* (La Ragione *et al.*, 2004), *E. coli* (Edens *et al.*, 1997; La Ragione *et al.*, 2004) and *Salmonella* (Zhang *et al.*, 2007; Carter *et al.*, 2017; La Ragione *et al.*, 2004) from poultry. Experimental challenge studies using the oral administration of viable bacteria showed that undefined complex cultures, particularly those derived directly from chicken intestines, were highly effective (Cameron and Carter, 1992). However, the use of such undefined agents derived from other birds brings a risk that pathogens may be part of the culture, resulting in the inadvertent transmission of disease agents (Jin *et al.*, 1998). This risk can be mitigated by the use of monocultures or well-defined multicultures to overcome this possible disadvantage (Reuter, 2001).

Following the 2006 EU ban on the sub-therapeutic use of antibiotics, one disease that rapidly became a target for alternative intervention strategies was avian intestinal spirochaetosis (AIS) (Collins *et al.*, 2009). AIS is an avian disease arising from the colonization of the caeca and colon of birds by the anaerobic spirochaete *Brachyspira* (McLaren *et al.*, 1997). AIS is particularly problematic for layer and broiler breeder flocks, due to the negative effect of *Brachyspira* colonization on the onset of lay, egg weights, diarrhoea occurrence, feed conversion efficiency and productivity of ovaries (Feberwee *et al.*, 2008). Of the seven documented species, three are considered pathogenic in poultry: *Brachyspira alvinipulli*, *Brachyspira intermedia* and *Brachyspira pilosicoli*. The prevalence of *Brachyspira* in free-range flocks in the UK has been estimated to be as high as 90% (Burch, 2010).

Colonization by *B. pilosicoli* has been characterized by an ability of the organism to form end-on attachments to and invasion of the intestinal epithelia, forming dense fringes that penetrate between enterocytes, leading to inflammatory responses (Feberwee *et al.*, 2008). This trait of *B. pilosicoli* has been exploited by researchers exploring non-antibiotic routes to combat AIS: among others, Se *et al.* (2008) initially demonstrated *in vitro* that *Lactobacillus johnsonii* exhibited antimicrobial activity against *B. pilosicoli* and *Brachyspira hyodysenteriae* through the production of hydrogen peroxide and a proteinaceous antimicrobial compound. Following these findings, further investigations showed that *Lactobacillus reuteri* LM1 and *Lactobacillus salivarius* LM2 inhibited the motility, growth and adherence of *B. pilosicoli* B2904 *in vitro*, even after the lactobacilli had been heat-killed or removed from culture to leave cell-free supernatant (Mappley *et al.*, 2011).

Subsequent *in vivo* studies more specifically investigated competitive exclusion of *L. reuteri* LM1 against AIS in 18-week-old pullets experimentally challenged via oral gavage with *B. pilosicoli* B2904 (Mappley *et al.*, 2013). These *in vivo* studies showed *L. reuteri* LM1 to be an ideal probiotic candidate for intervention against AIS due to the rapid growth and robust nature of lactobacilli compared with the slow-growing, fastidious *Brachyspira*. Mappley *et al.* (2013) also examined the potential for *L. reuteri* LM1 to competitively exclude *B. pilosicoli* B2904 and similarly found the probiotic was effective against AIS in laying hens.

Salmonella infections in poultry are also a particular focus for researchers owing to the dual concerns over the impact of infection on both the infected birds themselves and also humans consuming or handling infected meat or eggs. Poultry may be affected by many different strains of *Salmonella*, often without any clinical or performance effect in the birds. The likelihood of illness in poultry depends on a number of factors, including the *Salmonella* serotype and the age and general health of the bird. Poultry are most vulnerable to disease and death from *Salmonella* infection in the first two weeks post hatch, but infection in older birds such as breeding birds may lead to low hatchability with increased embryo mortality and increased mortality among newly hatched chickens. The most common infection routes for poultry are through contaminated feed and mixing of infected animals or eggs, but also through contact with a contaminated environment. The extensive *Salmonella* control programmes for poultry that exist globally consider routes of infection and are based on biosecurity prevention and extensive surveillance to prevent infection.

Interest in strains of *L. reuteri* as a probiotic defence system has extended over time to include addressing other infectious diseases such as *Salmonella* Enteritidis (SE), (Shi *et al.*, 2019), leading to an increasing body of evidence that differing strains of *L. reuteri* have potential applications as a probiotic agents due to their specific antibacterial activities against, for example, SE, including SE growth prevention, inhibition of gene expression of SE genes involved in adhesion and invasion, virulence and cell membrane integrity and inhibition of SE biofilm formation and bacterial motility.

The genus *Lactobacillus* has been a particularly rich source of candidate probiotic bacterial strains, probably in part as some are permanent members of the intestinal commensal microflora (Kullen and Klaenhammer 1999; Reid and Burton, 2002). In addition to the studies examining the use of poultry-derived bacterial strains as candidate probiotics acting against single pathogenic species, the potential for *Lactobacillus* strains to act as competitive exclusion agents against multiple bacterial pathogens in poultry has been considered. The capacity for the use of *Lactobacillus johnsonii* FI9785 as a competitive exclusion agent to reduce the colonization and shedding of *Salmonella enterica* serotype Enteritidis, *E. coli* O78:K80 and *Clostridium perfringens* in poultry was examined by La Ragione *et al.* (2004). The study demonstrated that *L. johnsonii* (FI9785) given as an oral bolus to poultry was effective in fully suppressing colonization by *C. perfringens* and showed some ability to interfere with the colonization and persistence of *E. coli*, but was ineffective against SE. This led to the suggestion that probiotic strains that had showed efficacy against one or more pathogenic strains of bacteria in poultry could be used in combination with other compatible strains to extend the range of activity against pathogens.

2.10 Use of Prebiotics to Enhance Healthy Gut Microbial Communities in Poultry

In addition to introducing beneficial bacteria to the gastrointestinal tract, the microbial population of the gastrointestinal tract may also be positively manipulated by introducing a prebiotic: a non-digestible dietary compound that selectively stimulates the growth and/or activity of one or a limited number of bacteria (Gibson and Roberfroid, 1995). Therefore the primary mechanism by which prebiotics support host animal intestinal health is mediated through beneficial bacteria via competitive exclusion of pathogens and generation of butyrate as an energy source for the intestinal epithelial cells (Tzortzis *et al.*, 2005). However, there is a growing body of evidence that glycoconjugates containing galactooligosaccharides (GOSs) and other oligosaccharides may also have the capacity to directly inhibit the attachment of pathogens to host cell surfaces by acting as competitive receptors (Arslanoglu *et al.*, 2007).

Salmonella Typhimurium has become a particular focus for developing non-antibiotic control methods owing to the emergence of antibiotic resistance (Carlson *et al.*, 2002) in *Salmonella* spp. Among the various oligosaccharides considered as candidate prebiotics, the capacity of some forms of GOS to reduce invasion and adhesion has has attracted considerable interest (Tzortzis *et al.*, 2005). A murine study by Searle *et al.* (2009) evaluated the efficacy of a prebiotic mixture containing GOS in mitigating against the adhesion and invasion of *S.* Typhimurium through a blend of *in vitro* and *in vivo* models. In tissue culture adherence and invasion assays with HT-29-16E cells, the presence of the prebiotic mix significantly reduced the invasion of *S.* Typhimurium. Similarly, in murine ligated ileal gut loops, the presence of the prebiotic mix prevented colonization and the associated pathology of *S.* Typhimurium. *In vivo* data using oral delivery of the prebiotic via a mouse model prior to challenge with *S.* Typhimurium showed significant reductions in colonization in the five organs sampled. Collectively, the results indicate that some GOS prebiotic mixes have the capacity to significantly reduce the colonization and pathology associated with *S.* Typhimurium infection in a murine model system. However, the study was not able to define whether any of the positive effects associated with GOS were directly through active binding of GOS to *S.* Typhimurium.

A subsequent study by the same group of researchers further investigated whether the protection conferred by the GOS mixture was impacting through direct action (Searle *et al.*, 2010). 3D culturing of HT-29-16E cells showed that the mixture significantly reduced the invasion of *S.* Typhimurium into these cells. This provides further indication that GOS may offer direct protective effects against *S.* Typhimurium by acting as a receptor mimic or modulating mucus production by host cells but studies are still required to elucidate the exact mechanisms of GOS action.

2.11 Use of Plant-derived Compounds to Combat Antibiotic Resistance

The use of bacterial strains with beneficial traits as probiotic feed additives offers substantial benefits to poultry health and the quest for alternative approaches to antibiotic use has created a broad vista of research areas, with some each focusing on very specific routes to combat the increasing levels of AMR. A group of six pathogenic bacteria known as the ESKAPE pathogens (*Enterococcus faecium, Staphylococcus aureus, Klebsiella pneumoniae, Acinetobacter baumannii, Pseudomonas aeruginosa* and *Enterobacter* species) are particularly urgent targets for non-antibiotic-based solutions as they harbour multiple genes associated with AMR (Rice, 2008).

A particularly problematic instance of AMR surrounds *P. aeruginosa*, a pathogenic bacterial species frequently associated severe and difficult-to-treat infections, such as wounds, burns and respiratory tract and bloodstream infections (Assael, 2011). Historically *P. aeruginosa* infections were effectively treated with aztreonam, but *P. aeruginosa* is now viewed by the WHO as a critically important threat to human health as the organism has become intrinsically resistant to antibiotic treatment.

Aztreonam is hydrolysed by extended-spectrum β-lactamases (ESBLs) and carbapenemases, and unfortunately, as many strains of *P. aeruginosa* possessing metallo-β-lactamases also carry other β-lactamases, aztreonam monotherapy is not always a valid treatment option (Jorth *et al.*, 2017). Therefore, research is underway to develop partnering substances, such as the β-lactamase inhibitor avibactam, for use alongside aztreonam (Marshall *et al.*, 2017). Another option to restore the activity of aztreonam against bacterial strains with multiple resistance mechanisms would be to use polyphenols.

Polyphenols such as epigallocatechin gallate (EGCG) are biologically active compounds shown to have antibacterial activity alone against many species of bacteria, including *Acinetobacter baumannii* and *Stenotrophomonas maltophilia* (Steinmann *et al.*, 2013). The potential for use of theaflavin-epicatechin combinations versus multidrug-resistant *Acinetobacter baumannii* has been recently explored using the greater wax moth larvae, *Galleria mellonella*, as an *in vivo* model for bacteraemia, because *G. mellonella* has been shown to have an innate immune system comparable with that of mammals (Betts *et al.*, 2011). The study examined the *in vitro* kill kinetics and the *in vivo* toxicity and efficacy of theaflavin–epicatechin combinations against multidrug-resistant strains of *A. baumannii* and found that the theaflavin–epicatechin combination was not only synergistic but also bactericidal. The study also highlighted a lack of toxicity to polyphenols even when administrated at high concentrations to *G. mellonella*, but these observations might be due an evolved dietary tolerance of phenolic and polyphenolic compounds so potential toxicity in mammalian models is necessary. A subsequent study by Betts *et al.* (2019) investigated whether EGCG could restore the *in vitro* and *in vivo* activity of aztreonam against multidrug-resistant strains of *P. aeruginosa* and found not only that synergy exists between aztreonam and EGCG against multidrug-resistant strains of *P. aeruginosa* but also that EGCG could restore the antibacterial activity of aztreonam to concentrations below the European Committee on Antimicrobial Susceptibility Testing (EUCAST) recommended susceptibility breakpoint for *P. aeruginosa*, potentially expanding and extending its useful therapeutic lifespan.

2.12 Use of PhotoCORMS Against Avian Pathogenic *E. coli*-induced Disease

PhotoCORMS are UV-activated carbon monoxide (CO) releasing molecules. They are mostly metal carbonyl complexes that act as anti-inflammatory and cytoprotective molecules. Differing photoCORM molecules display variable kinetics and conditions for CO release, allowing specific selection for particular environments. The capacity of photoCORMS to target critical oxidases is now being harnessed to address the growing global concerns over AMR in APEC. While in a healthy bird, APECs make up part of the normal avian intestinal flora, an imbalance in the intestinal microflora can cause avian colibacillosis; this is a clinically serious condition in poultry caused by APEC involving respiratory and systemic disease, leading to negative impacts on poultry production welfare and economics costs on the poultry industry worldwide (La Ragione and Woodward, 2002). CORMs are an emerging group of drug candidates with potential for therapeutic use against Gram-negative bacteria including *E. coli*.

Betts *et al.* (2017) demonstrated, both *in vitro* and *in vivo*, the antibacterial action of a manganese(I) carbonyl complex CORM against APEC and that activity of the CORM significantly increases when used in conjunction with colistin. The addition of the colistin increased the antibacterial activity of the manganese(I) carbonyl complex CORM against multidrug-resistant strains of APEC by permeabilizing the APEC outer membrane to facilitate entry of the metal complex.

2.13 Conclusion

AMR remains a global challenge; however, significant progress has been made to understand the drivers of AMR. It is clear that reducing the use of antibiotics in the poultry sector is important, but AMR is multifactorial and thus needs to be tackled via a multi-sectorial approach.

Ongoing research is being conducted in commercial settings with the support of poultry veterinarians and in world-class research facilities in a concerted effort to ensure poultry intestinal health is maximized to improve health, welfare and productivity. However, collaboration between medical, veterinary and other allied professions is key to transferring emerging research rapidly to application in the field, especially regarding AMR.

Through an increased understanding of the makeup of the gut poultry microflora it may be possible to gain a detailed understanding of the pathobiology of diseases and develop novel interventions. Many interventions, including prebiotics, probiotics, phytochemicals and CORMS, have the potential to modulate the gut flora. The novel interventions may be employed to control poultry pathogens and will, hopefully, become more widely used as our understanding of their modes of action increases. Beyond this, the recent advances in screening and characterizing the metabolome and microflora/microbiome signatures can be used to monitor disease status, reduce AMR and improve the health and welfare of poultry. Hopefully, such tools will in the future also help the poultry sector to continue to play a vital role in reducing AMR.

References

Aidara-Kane, A., Angulo, F.J., Conly, J., Minato, Y., Silbergeld, E.K. *et al.* (2018) World Health Organization (WHO) guidelines on use of medically important antimicrobials in food-producing animals. *Antimicrobial Resistance and Infection Control* 7, 1–8. doi: 10.1186/S13756-017-0294-9/TABLES/3.

Anderson, D.B., McCracken, V.J., Aminov, R.I., Simpson, J.M., Mackie, R.I. *et al.* (2000) Gut microbiology and growth-promoting antibiotics in swine. *Pig News and Information* 20, 1115–1122.

Antharam, V.C., Li, E.C., Ishmael, A., Sharma, A., Mai, V. *et al.* (2013) Intestinal dysbiosis and depletion of butyrogenic bacteria in *Clostridium difficile* infection and nosocomial diarrhea. *Journal of Clinical Microbiology* 51, 2884–2892. doi: 10.1128/JCM.00845-13/SUPPL_FILE/ZJM999092753SO1.PDF.

Arslanoglu, S., Moro, G.E. and Boehm, G. (2007) Early supplementation of prebiotic oligosaccharides protects formula-fed infants against infections during the first 6 months of life. *The Journal of Nutrition* 137, 2420–2424. doi: 10.1093/JN/137.11.2420.

Assael, B.M. (2011) Aztreonam inhalation solution for suppressive treatment of chronic *Pseudomonas aeruginosa* lung infection in cystic fibrosis. *Expert Review of Anti-infective Therapy* 9, 967–973. doi: 10.1586/ERI.11.131.

Bennett, C.E., Thomas, R., Williams, M., Zalasiewicz, J., Edgeworth, M. *et al.* (2018) The broiler chicken as a signal of a human reconfigured biosphere. *Royal Society Open Science* 5. doi: 10.1098/RSOS.180325

Betts, J.W., Kelly, S.M. and Haswell, S.J. (2011) Antibacterial effects of theaflavin and synergy with epicatechin against clinical isolates of *Acinetobacter baumannii* and *Stenotrophomonas maltophilia*. *International Journal of Antimicrobial Agents* 38, 421–425 doi: 10.1016/J.IJANTIMICAG.2011.07.006.

Betts, J.W., Nagel, C., Schatzschneider, U., Poole, R. and La Ragione, R.M. (2017) Antimicrobial activity of carbon monoxide-releasing molecule [Mn(CO)3(tpa-κ3N)]Br versus multidrug-resistant isolates of Avian Pathogenic *Escherichia coli* and its synergy with colistin. *PLOS ONE* 12, e0186359. doi: 10.1371/JOURNAL.PONE.0186359.

Betts, J.W., Hornsey, M. and La Ragione, R.M. (2018) Novel antibacterials: alternatives to traditional antibiotics. *Advances in Microbial Physiology* 73, 123–169. doi: 10.1016/BS.AMPBS.2018.06.001.

Betts, J.W., Hornsey, M., Higgins, P.G., Lucassen, K., Wille, J. *et al.* (2019) Restoring the activity of the antibiotic aztreonam using the polyphenol epigallocatechin gallate (EGCG) against multidrug-resistant clinical isolates of *Pseudomonas aeruginosa*. *Journal of Medical Microbiology* 68, 1552–1559. doi: 10.1099/JMM.0.001060/CITE/REFWORKS.

Bovenhuis, H., Berghof, T.V.L., Visker, M.H.P.W., Arts, J.A.J., Visscher, J. *et al.* (2022) Divergent selection for natural antibodies in poultry in the presence of a major gene. *Genetics Selection Evolution* 54, 1–13. doi: 10.1186/S12711-022-00715-9/TABLES/5

Brummermann, M. and Braun, E.J. (1995) Effect of salt and water balance on colonic motility of white leghorn roosters. *American Journal of Physiology* 268, 690–698. doi: 10.1152/AJPREGU.1995.268.3.R690

Burch, D. (2010) Egg production drops from *Brachyspira*. *Poultry World*. Available at: https://www.poultryworld.net/poultry/egg-production-drops-from-brachyspira/ (accessed 27 January 2023).

Cameron, D.M. and Carter, J.N. (1992) Evaluation of the efficacy of Broilact® in preventing infection of broiler chicks with *Salmonella* enteritidis PT4. *International Journal of Food Microbiology* 15, 319–326. doi: 10.1016/0168-1605(92)90065-B

Card, R.M., Cawthraw, S.A., Nunez-Garcia, J., Ellis, R.J., Kay, G. *et al.* (2017) An *in vitro* chicken gut model demonstrates transfer of a multidrug resistance plasmid from Salmonella to commensal *Escherichia coli*. *MBio*, 8(4). doi: 10.1128/MBIO.00777-17/SUPPL_FILE/MBO003173381ST7.XLSX.

Carlson, S.A., Stoffregen, W.C. and Bolin, S.R. (2002) Abomasitis associated with multiple antibiotic resistant *Salmonella* enterica serotype Typhimurium phagetype DT104. *Veterinary Microbiology* 85, 233–240. doi: 10.1016/S0378-1135(01)00508-9.

Carter, A., Adams, M., La Ragione, R.M. and Woodward, M.J. (2017) Colonisation of poultry by *Salmonella Enteritidis* S1400 is reduced by combined administration of *Lactobacillus salivarius* 59 and *Enterococcus faecium* PXN-33. *Veterinary Microbiology* 199, 100–107.

Casewell, M., Friis, C., Marco, E., McMullin, P. and Phillips, I. (2003) The European ban on growth-promoting antibiotics and emerging consequences for human and animal health. *Journal of Antimicrobial Chemotherapy* 52(2), 159–161. doi: 10.1093/JAC/DKG313.

Castanon, J.I.R. (2007) History of the use of antibiotic as growth promoters in European poultry feeds. *Poultry Science* 86, 2466–2471. https://doi.org/10.3382/PS.2007-00249.

Chambers, M.A., Graham, S.P. and La Ragione, R.M. (2016) Challenges in veterinary vaccine development and immunization. *Methods in Molecular Biology* 1404, 3–35. doi: 10.1007/978-1-4939-3389-1_1/TABLES/1.

Chattopadhyay, M.K. (2014) Use of antibiotics as feed additives: A burning question. *Frontiers in Microbiology* 5, 334. doi: 10.3389/FMICB.2014.00334/BIBTEX.

Chrzastek, K., Leng, J., Zakaria, M.K., Bialy, D., La Ragione, R. and Shelton, H. (2021) Low pathogenic avian influenza virus infection retards colon microbiota diversification in two different chicken lines. *Animal Microbiome* 3, 1–15. doi: 10.1186/S42523-021-00128-X.

Collins, J.W., La Ragione, R.M., Woodward, M.J. and Searle, L.E.J. (2009) Application of pre- and probiotics in livestock. In: D. Charalampopoulos and R.A. Rastall (eds). *Prebiotics and Probiotics Science and Technology* Springer, Berlin, pp. 57–88.

Cui, N., Huang, X., Kong, Z., Huang, Y., Huang, Q. *et al.* (2018) Newcastle disease virus infection interferes with the formation of intestinal microflora in newly hatched specific-pathogen-free chicks. *Frontiers in Microbiology* 9, 900. doi: 10.3389/FMICB.2018.00900/BIBTEX.

Davies, R. and Wales, A. (2019) Antimicrobial resistance on farms: a review including biosecurity and the potential role of disinfectants in resistance selection. *Comprehensive Reviews in Food Science and Food Safety* 18, 753–774. doi: 10.1111/1541-4337.12438.

De Maesschalck, C., Eeckhaut V., Maertens L., De Lange L., Marchal L. *et al.* (2015) Effects of xylo-oligosaccharides on broiler chicken performance and microbiota. *Applied Environmental Microbiology* 81, 5880–5888. doi: 10.1128/AEM.01616-15.

Edens, F.W., Qureshi, R.A., Parkhurst, C.R., Qureshi, M.A., Havenstein, G.B. and Casas, I.A. (1997) Characterization of two *Escherichia coli* isolates associated with poult enteritis and mortality syndrome. *Poultry Science* 76, 1665–1673. doi: 10.1093/PS/76.12.1665.

FDA (2012) Guidance for Industry: The Judicious Use of Medically Important Antimicrobial Drugs in Food-Producing Animals. Available at: https://www.fda.gov/media/79140/download (accessed 27 March 2023).

Feberwee, A., Hampson, D.J., Phillips, N.D., La, T., van der Heijden, H.M.J.F. *et al.* (2008) Identification of *Brachyspira hyodysenteriae* and other pathogenic *Brachyspira* species in chickens from laying flocks with diarrhea or reduced production or both. *Journal of Clinical Microbiology* 46, 593. https://doi.org/10.1128/JCM.01829-07.

Gibson, G.R. and Roberfroid, M.B. (1995) Dietary modulation of the human colonic microbiota: introducing the concept of prebiotics. *The Journal of Nutrition* 125, 1401–1412. doi: 10.1093/JN/125.6.1401.

Gilroy, R., Ravi, A., Getino, M., Pursley, I., Horton, D.L. *et al.* (2021) Extensive microbial diversity within the chicken gut microbiome revealed by metagenomics and culture. *PeerJ*, 9, e10941. doi: 10.7717/PEERJ.10941/SUPP-2.

Glendinning, L., Stewart, R.D., Pallen, M.J., Watson, K.A. and Watson, M. (2020) Assembly of hundreds of novel bacterial genomes from the chicken caecum. *Genome Biology* 21, 1–16. doi: 10.1186/S13059-020-1947-1/FIGURES/8.

Hillmann, B., Al-Ghalith, G.A., Shields-Cutler, R.R., Zhu, Q., Gohl, D.M. *et al.* (2018) Evaluating the Information Content of Shallow Shotgun Metagenomics. *MSystems* 3(6). doi: 10.1128/MSYSTEMS.00069-18/SUPPL_FILE/SYS005182277ST2.TXT.

Huang, P., Zhang, Y., Xiao, K., Jiang, F., Wang, H. *et al.* (2018) The chicken gut metagenome and the modulatory effects of plant-derived benzylisoquinoline alkaloids. *Microbiome* 6(1), 211. doi: 10.1186/S40168-018-0590-5/FIGURES/5.

Jin, L.Z., Ho, Y.W., Abdullah, N. and Jalaludin, S. (1998) Growth performance, intestinal microbial populations, and serum cholesterol of broilers fed diets containing *Lactobacillus* cultures. *Poultry Science* 77, 1259–1265. doi: 10.1093/PS/77.9.1259.

Jorth, P., McLean, K., Ratjen, A., Secor, P.R., Bautista, G.E. *et al.* (2017) Evolved aztreonam resistance is multifactorial and can produce hypervirulence in *Pseudomonas aeruginosa*. *MBio* 8(5). doi: 10.1128/MBIO.00517-17/SUPPL_FILE/MBO005173556SF3.PDF.

Jukes, T.H. and Williams, W.L. (1953) Nutritional effects of antibiotics. *Pharmacological Reviews* 5, 381–420.

Korver, D.R. (2022) Intestinal nutrition: role of vitamins and biofactors and gaps of knowledge. *Poultry Science* 101, 101665. doi: 10.1016/J.PSJ.2021.101665.

Kullen, M.J. and Klaenhammer, T.R. (1999) Identification of the pH-inducible, proton-translocating F1F0-ATPase (atpBEFHAGDC) operon of *Lactobacillus acidophilus* by differential display: gene structure, cloning and characterization. *Molecular Microbiology* 33, 1152–1161. doi: 10.1046/J.1365-2958.1999.01557.X.

La Ragione, R.M. and Woodward, M.J. (2002) Virulence factors of *Escherichia coli* serotypes associated with avian colisepticaemia. *Research in Veterinary Science* 73, 27–35. doi: 10.1016/S0034-5288(02)00075-9.

La Ragione, R.M., Narbad, A., Gasson, M.J. and Woodward, M.J. (2004) *In vivo* characterization of *Lactobacillus johnsonii* FI9785 for use as a defined competitive exclusion agent against bacterial pathogens in poultry. *Letters in Applied Microbiology* 38, 197–205. doi: 10.1111/J.1472-765X.2004.01474.X.

Lagerstrom, K.M. and Hadly, E.A. (2021) The under-investigated wild side of *Escherichia coli*: genetic diversity, pathogenicity and antimicrobial resistance in wild animals. *Proceedings of the Royal Society B* 288. https://doi.org/10.1098/RSPB.2021.0399.

Landers, T.F., Cohen, B., Wittum, T.E. and Larson, E.L. (2012) A Review of Antibiotic Use in Food Animals: Perspective, Policy, and Potential. *Public Health Reports* 127, 4–22. doi: 10.1177/003335491212700103.

Larsen, P.B. (2002) Consequences of terminating AGP use for pig health and usage of antimicrobials for therapy and prophylaxis. *Abstracts of the International Invitational Symposium: Beyond Antibiotic Growth Promoters in Food Animal Production*, Foulum, Denmark, 2002, 67–71.

Le Roy, C.I., Woodward, M.J., Ellis, R.J., La Ragione, R.M. and Claus, S.P. (2019) Antibiotic treatment triggers gut dysbiosis and modulates metabolism in a chicken model of gastro-intestinal infection. *BMC Veterinary Research* 15, 1–13. doi: 10.1186/S12917-018-1761-0/FIGURES/5.

Lea, H., Spring, P., Taylor-Pickard, J. and Burton, E. (2013) A natural carbohydrate fraction ActigenTM from Saccharomyces cerevisiae cell wall: effects on goblet cells, gut morphology and performance of broiler chickens. *Journal of Applied Animal Nutrition* 1, 1–7. doi: 10.1017/JAN.2013.6.

Leng, J., Walton, G., Swann, J., Darby, A., La Ragione, R. and Proudman, C. (2019) 'Bowel on the bench': Proof of concept of a three-stage, *in vitro* fermentation model of the equine large intestine. *Applied and Environmental Microbiology* 86(1). doi: 10.1128/AEM.02093-19/SUPPL_FILE/AEM.02093-19-S0001.PDF.

Lhermie, G., La Ragione, R.M., Weese, J.S., Olsen, J.E., Christensen, J.P. and Guardabassi, L (2020) Indications fort the use of the highest priority critically important antimicrobials in the veterinary sector. *Journal of Antimicrobial Chemotherapy* 2. doi: 10.1093/jac/dkaa104.

Long, J.S., Idoko-Akoh, A., Mistry, B., Goldhill, D., Staller, E. *et al.* (2019) Species specific differences in use of ANP32 proteins by influenza A virus. *ELife* 8, 696–697. doi: 10.7554/ELIFE.45066.001.

Magnúsdóttir, S., Ravcheev, D., de Crécy-Lagard, V. and Thiele, I. (2015) Systematic genome assessment of B-vitamin biosynthesis suggests cooperation among gut microbes. *Frontiers in Genetics* 6, 148. doi: 10.3389/FGENE.2015.00148/ABSTRACT.

Mappley, L.J., Tchórzewska, M.A., Cooley, W.A., Woodward, M.J. and La Ragione, R.M. (2011) Lactobacilli antagonize the growth, motility, and adherence of *Brachyspira pilosicoli*:

A potential intervention against avian intestinal spirochetosis. *Applied and Environmental Microbiology* 77, 5402–5411. doi: 10.1128/AEM.00185-11/ASSET/C3478862-08B2-4FD3-B61E-F78C794473F8/ASSETS/GRAPHIC/ZAM9991023330005.JPEG.

Mappley, L.J., Tchórzewska, M.A., Nunez, A., Woodward, M.J., Bramley, P.M. and La Ragione, R.M. (2013) Oral treatment of chickens with *Lactobacillus reuteri* LM1 reduces *Brachyspira pilosicoli*-induced pathology. *Journal of Medical Microbiology* 62, 287–296.

Marshall, S., Hujer, A.M., Rojas, L.J., Papp-Wallace, K.M., Humphries, R.M. *et al.* (2017) Can ceftazidime-avibactam and aztreonam overcome β-lactam resistance conferred by metallo-β-lactamases in Enterobacteriaceae? *Antimicrobial Agents and Chemotherapy* 61(4).

Mclaren, A.J., Trott, D.J., Swayne, D.E., Oxberry, S.L. and Hampson, D.J. (1997) Genetic and phenotypic characterization of intestinal spirochetes colonizing chickens and allocation of known pathogenic isolates to three distinct genetic groups. *Journal of Clinical Microbiology* 35, 412–417.

Medvecky, M., Cejkova, D., Polansky, O., Karasova, D., Kubasova, T. *et al.* (2018) Whole genome sequencing and function prediction of 133 gut anaerobes isolated from chicken caecum in pure cultures. *BMC Genomics* 19, 1–15. https://doi.org/10.1186/S12864-018-4959-4/FIGURES/5

Milner, K.C. and Shaffer, M.F. (1952) Bacteriologic studies of experimental *Salmonella* infections in chicks. *The Journal of Infectious Diseases* 90, 81–96. Available at: https://www.jstor.org/stable/30094006 (accessed 27 March 2023).

Murray, J.M.D., McMullin, P., Handel, I. and Hastie, P.M. (2014) Comparison of intestinal contents from different regions of the equine gastrointestinal tract as inocula for use in an *in vitro* gas production technique. *Animal Feed Science and Technology* 187, 98–103. doi: 10.1016/J.ANIFEEDSCI.2013.10.005.

Nichols, G.L., Richardson, J.F., Sheppard, S.K., Lane, C. and Sarran, C. (2012) *Campylobacter* epidemiology: a descriptive study reviewing 1 million cases in England and Wales between 1989 and 2011. *BMJ Open* 2, e001179. doi: 10.1136/BMJOPEN-2012-001179.

NRC (1999) The Use of Drugs in Food Animals. Available at: https://doi.org/10.17226/5137 (accessed 27 March 2023).

Nurmi, E. and Rantala, M. (1973) New aspects of Salmonella infection in broiler production. *Nature* 241, 210–211.

OECD/FAO (2021) Table C.28.1 – Poultry meat projections: Production and trade, *OECD-FAO Agricultural Outlook 2021-2030*, OECD Publishing, Paris. Available at: https://doi.org/10.1787/aaefa116-en (accessed 27 March 2023).

Qu, A., Brulc, J.M., Wilson, M.K., Law, B.F., Theoret, J.R. *et al.* (2008) Comparative metagenomics reveals host specific metavirulomes and horizontal gene transfer elements in the chicken cecum microbiome. *PLOS ONE* 3, e2945. doi: 10.1371/JOURNAL.PONE.0002945.

Reid, G. and Burton, J. (2002) Use of *Lactobacillus* to prevent infection by pathogenic bacteria. *Microbes and Infection* 4, 319–324. doi: 10.1016/S1286-4579(02)01544-7.

Reuter, G. (2001) The *Lactobacillus* and *Bifidobacterium* microflora of the human intestine: composition and succession. *Current Issues in Intestinal Microbiology* 2, 43–53.

Rice, L.B. (2008) Federal funding for the study of antimicrobial resistance in nosocomial pathogens: No ESKAPE. *The Journal of Infectious Diseases* 197, 1079–1081. doi: 10.1086/533452.

Ritter, J.M. (2008) Drug regulation & therapeutic efficacy. *British Journal of Clinical Pharmacology*, 65, 801–802. doi: 10.1111/J.1365-2125.2008.03208.X.

Rychlik, I. (2020) Composition and function of chicken gut microbiota. *Animals,* 10, 103. doi: 10.3390/ANI10010103.

Sakaridis, I., Ellis, R.J., Cawthraw, S.A., van Vliet, A.H.M., Stekel, D.J. *et al.* (2018) Investigating the association between the caecal microbiomes of broilers and *Campylobacter* burden. *Frontiers in Microbiology* 9, 927. doi: 10.3389/FMICB.2018.00927/BIBTEX.

Se, C.-Y., Yong, F.-F., Tan, H.M., Yeo, W.M. and Yeow-Lim Teo, A. (2008) Broad-spectrum antibacterial and antifungal activity of *Lactobacillus johnsonii* D115. U.S. patent 20080299098.

Searle, L.E.J., Best, A., Nunez, A., Salguero, F.J., Johnson, L. *et al.* (2009) A mixture containing galactooligosaccharide, produced by the enzymic activity of *Bifidobacterium bifidum*, reduces *Salmonella* enterica serovar typhimurium infection in mice. *Journal of Medical Microbiology* 58, 37–48. doi: .1099/JMM.0.004390-0/CITE/REFWORKS.

Searle, L.E.J., Cooley, W.A., Jones, G., Nunez, A., Crudgington, B. *et al.* (2010) Purified galactooligosaccharide, derived from a mixture produced by the enzymic activity of *Bifidobacterium bifidum*, reduces *Salmonella* enterica serovar Typhimurium adhesion and invasion *in vitro* and *in vivo*. *Journal of Medical Microbiology* 59, 1428–1439. doi: 10.1099/JMM.0.022780-0/CITE/REFWORKS.

Sergeant, M.J., Constantinidou, C., Cogan, T.A., Bedford, M.R., Penn, C.W. and Pallen, M.J. (2014) Extensive microbial and functional diversity within the chicken cecal microbiome. *PLOS ONE* 9, e91941. doi: 10.1371/JOURNAL.PONE.0091941.

Shang, Y., Kumar, S., Oakley, B. and Kim, W.K. (2018) Chicken gut microbiota: Importance and detection technology. *Frontiers in Veterinary Science* 5, 254. ?doi: 10.3389/FVETS.2018.00254/BIBTEX.

Shi, S., Qi, Z., Sheng, T., Tu, J., Shao, Y. and Qi, K. (2019) Antagonistic trait of *Lactobacillus reuteri* S5 against *Salmonella enteritidis* and assessment of its potential probiotic characteristics. *Microbial Pathogenesis* 137, 103773. doi: 10.1016/J.MICPATH.2019.103773.

Spor, A., Koren, O. and Ley, R. (2011) Unravelling the effects of the environment and host genotype on the gut microbiome. *Nature Reviews Microbiology* 9, 279–290. doi: 10.1038/NRMICRO2540.

Steinmann, J., Buer, J., Pietschmann, T. and Steinmann, E. (2013) Anti-infective properties of epigallocatechin-3-gallate (EGCG), a component of green tea. *British Journal of Pharmacology* 168, 1059–1073. https://doi.org/10.1111/BPH.12009.

Swann, M. (1969) The use of antibiotics in animal husbandry and veterinary medicine (Swann report). HMSO, London.

Thomson, N.M., Gilroy, R., Getino, M., Foster-Nyarko, E., van Vliet, A.H.M. *et al.* (2022) Remarkable genomic diversity among *Escherichia* isolates recovered from healthy chickens. *Peer J* 10, e12935. doi: 10.7717/PEERJ.12935/SUPP-10.

Tzortzis, G., Goulas, A.K., Gee, J.M. and Gibson, G.R. (2005) A novel galactooligosaccharide mixture increases the Bifidobacterial population numbers in a continuous *in vitro* fermentation system and in the proximal colonic contents of pigs *in vivo*. *The Journal of Nutrition* 135, 1726–1731. doi: 10.1093/JN/135.7.1726.

USDA (2022) *Campylobacter* in Poultry. Available at: https://www.nifa.usda.gov/sites/default/files/resource/Campylobacter%20in%20Poultry.pdf (accessed 19 January 2023).

Vanderpool, C., Yan, F. and Polk, D.B. (2008) Mechanisms of probiotic action: implications for therapeutic applications in inflammatory bowel diseases. *Inflammatory Bowel Disease*14, 1585–1596. doi: 10.1002/ibd.20525.

WHO (2017) Stop using antibiotics in healthy animals to prevent the spread of antibiotic resistance. Available at: https://www.who.int/news/item/07-11-2017-stop-using-antibiotics-in-healthy-animals-to-prevent-the-spread-of-antibiotic-resistance (accessed 27 March 2023).

WHO (2018) WHO report on surveillance of antibiotic consumption: 2016-2018 early implementation. Geneva: World Health Organization; 2018. Licence: CC BY-NC-SA 3.0 IGO Available at: https://www.who.int/publications/i/item/who-report-on-surveillance-of-antibiotic-consumption (accessed 27 March 2023).

Wierup, M. (2001) The Swedish experience of the 1986 year ban of antimicrobial growth promoters, with special reference to animal health, disease prevention, productivity, and usage of antimicrobials. *Microbial Drug Resistance* 7, 183–190. doi: 10.1089/10766290152045066.

Williams-Nguyen, J., Sallach, J.B., Bartelt-Hunt, S., Boxall, A.B., Durso, L.M. *et al.* (2016) Antibiotics and antibiotic resistance in agroecosystems: state of the science. *Journal of Environmental Quality* 45, 394–406. doi: 10.2134/jeq2015.07.0336.
Xu, P., Liu, P., Zhou, C., Shi, Y., Wu, Q. *et al.* (2019) A multi-omics study of chicken infected by nephropathogenic infectious bronchitis virus. *Viruses* 11, 1070. doi: 10.3390/V11111070.
Yitbarek, A., Taha-Abdelaziz, K., Hodgins, D.C., Read, L., Nagy, É. *et al.* (2018) Gut microbiota-mediated protection against influenza virus subtype H9N2 in chickens is associated with modulation of the innate responses. *Scientific Reports* 8, 1–12. doi: 10.1038/s41598-018-31613-0.
Zhang, G., Ma, L. and Doyle, M.P. (2007) Potential competitive exclusion bacteria from poultry inhibitory to *Campylobacter jejuni* and *Salmonella*. *Journal of Food Protection* 70, 867–873. doi: 10.4315/0362-028X-70.4.867.

CHAPTER 3
A Practical Approach to Managing Poultry Production Without Antibiotics

Reg Smith[1]* and Daniel Parker[2]

[1]*The Stables, Princethorpe, Warwickshire, UK;* [2]*Slate Hall, Metheringham, Lincolnshire, UK*

Abstract

To maintain and improve the long-term efficacy of antibiotics for both human and animal health, a global approach to responsible antibiotic stewardship requires targeted reduction in the use of all types of antibiotics in human medicine, food production and farming. This chapter describes practical approaches necessary to manage poultry production without the requirement for non-therapeutic antibiotic treatment. Poultry farming operations supported by poultry health professionals came together under the umbrella of the British Poultry Council (BPC) antibiotic stewardship scheme. Jointly, through practical application on commercial farms, key techniques of husbandry, hygiene and health management were identified that, if applied, collectively reduced the need for antibiotic intervention. Through this work, antibiotic usage within the UK poultry meat sector was reduced by 84% over a 4-year period. To achieve this reduction, the management processes and hygiene controls through the length of the production chain from breeding farms through hatcheries to commercial farms had to be controlled to minimize the bacterial loadings and insults to the bird's immune systems. Personnel who could have the greatest impact upon the success of antibiotic reduction strategies were those who cared for the birds on a daily basis.

3.1 Antibiotic Growth Promoters

The discovery of antibiotics by Fleming in the 1940s heralded an era whereby many common bacterial diseases affecting humans could be virtually eliminated

*Corresponding author: Reg@Princethorpepoultry.co.uk

DOI: 10.1079/9781800622746.0003

in many parts of the world. On farms in whaling and fishing fleets, as well as in processing plants and aquaculture operations, antibiotics were used to treat and prevent disease, increase feed conversion, and preserve food (Kirchhelle, 2018).

After World War II, the drive to feed the population harnessed emerging antibiotics. In the USA in 1948, Merck's Sulfaquinoxaline® was the first licensed product for inclusion in poultry feedstuffs. The performance benefits that could be achieved by using antibiotics led to a proliferation of products for use in livestock production. Initially used for the treatment of clinical disease, antibiotics started to be used to suppress sub-clinical disease issues. Frequently, usage was aimed at suppressing secondary bacterial infections following some other insult to the bird's immune system. The point came where they were effectively being used to promote growth. By the 1970s, the era of antibiotic growth promoters (AGPs) had truly arrived. Poultry, along with other types of farm livestock, received these AGPs in the feed ration for most or all their lives.

For clarity, the following definition of growth promotion has been used:

> Growth promotion refers to the use of substances to increase the rate of weight gain and or feed efficiency utilization in animals by other than purely nutritional means. (FAO, 2015)

It is the author's personal contention that the use of AGPs in boosting performance of a group of animals negated the labour and resource that had been previously required to care for sick animals. Farms could deliver acceptable performance when using AGPs, which, in their absence, may well have required a higher standard of hygiene and biosecurity and more rigorous management processes. Bluntly, livestock owners could reach performance criteria with less effort required than was previously possible.

Critics of livestock farming at the time claimed, rightly or wrongly, that AGPs masked sub-optimal on-farm management. As an active agriculturalist during this period, two factors allowed for a way of working that, when viewed through today's prism, is now generally thought to be poor antibiotic stewardship. First, the subject of antimicrobial resistance (AMR) was not spoken about at farm level. As such, farmers potentially lacked the level of knowledge that they have available to them today. Second, and just as importantly, the levels of independent on-farm scrutiny and accompanying audit processes common today were, at that time, absent to the greatest degree.

During the 1970s and 1980s AGPs were commonly included in poultry feeds. It is likely many individual farmers either did not realize or did not have reason to question the inclusion of such products. At that time, they were included in the bulk of feed fed to all livestock. Examples of some commonly used AGPs in poultry and pig production are shown in Table 3.1.

Interestingly, historically the use of AGPs was acceptable, but by the 1990s there was developing awareness of the need to place greater controls on antibiotic usage. For clarity, under today's analysis, two key issues existed with the then administration of AGPs. Firstly, the administration of AGPs existed without significant clinical investigation or identification of an actual disease. Effectively, it was prophylactic administration. Secondly, at least two of the

Table 3.1. Some commonly used AGP's in poultry and/or pigs.

Antibiotic	Class	HPCIA[a]
Avoparcin	Glycopeptide	Yes
Tylosin phosphate	Macrolide	Yes
Virginiamycin	Streptogramin	
Avilamycin	Orthosomycin	

[a]Highest priority critically important antibiotics according to the World Health Organization

AGPs listed above were from classes of antibiotics that are now classified by the World Health Organization (WHO) as belonging to the highest priority critically important antibiotics (HPCIAs) needed to maintain human health (WHO, 2016).

3.2 The Breaking Storm

In the mid-1990s, negative media coverage against intensive farming was common. The use of antibiotics in livestock production fuelled some of that coverage. Media reporting of the issue was either considered as fact or sensationalistic journalism, depending on your own viewpoint. Irrespectively, the subject of antibiotic usage in livestock production was high on the agenda for the poultry meat industry. Poultry meat was, and still is, one of the major proteins consumed within the UK; as such, it received a great deal of attention from concerned stakeholders.

A chain reaction of bans on all growth-promoting antibiotics was started in Sweden in 1986 and the ban on avoparcin and virginiamycin came into force in Denmark in 1995 and 1998. The European Union (EU) banned the use of avoparcin in 1997 and the four remaining antibiotics used for growth promotion in 1999 based on the 'Precautionary Principle'. These last four antibiotics were: bacitracin, spiramycin and tylosin (macrolides) and virginiamycin (a streptogramin combination) (Casewell *et al.*, 2003).

Meanwhile in the UK, in September 1999, one of the largest broiler processing companies (by slaughter numbers) led the way by announcing an immediate ban on the use of all growth-promoting antibiotics. Retail pressure resulted in the remaining industry players doing likewise in a short timeframe. The downstream risks of such a rapid removal soon played out. Bacterial overgrowth within the bird's gut, principally by *Clostridium perfringens* led to necrotic enteritis being commonplace. The full epidemiology of necrotic enteritis is beyond the scope of this chapter; however, it is a disease that impacts the normal function of the gut and, in worst case situations, can lead to the death of elements of the gastrointestinal tract and thereby of the bird. The impact of necrotic enteritis on bird performance included mortality levels of up to 25% and a loss of up to 20 points of feed efficiency. The abrupt removal of AGPs from the diet resulted in a general instability of the bird's intestinal microbiota and the consequences were loose droppings, wet bedding, loss of bird performance, morbidity and, in the worst cases, mortality. These personal performance observations were repeated across most of the commercial broiler operations within the UK.

The potential risks associated with the removal of AGPs, i.e. losing control of clostridia, and the effects that change would have on gut health were telegraphed by veterinarians and other animal health experts. The impacts of removing AGPs upon the health of the broiler flocks within the UK were clearly not a surprise to those working to support animal health. However, the author's personal perception was the degree to which health and welfare was impacted was far greater than expected. The resulting impairment to the gut health increasingly promoted an environment where coccidiosis added further insult. Unabated, the combination of heavy clostridial infections and elevated protozoal populations devastated physical and financial farm performance.

The immediate solution required a two-pronged attack: antibiotics to quell the clostridia and coccidiostats to support coccidiosis control. Antibiotics such as potassium penicillin or similar were used extensively and frequently used repetitively within one flock cycle. Meanwhile, ionophore coccidiostats became more popular in the treatment of coccidiosis. The irony of this chapter in the history of the broiler industry was that although a positive move to reduce the use of AGPs succeeded in lowering overall antibiotic usage, to deal with the ensuing health problems elevated levels of antibiotics were being used therapeutically and these antibiotics were generally those classes used more commonly in human medicine.

It was Benjamin Franklin who said 'out of adversity comes opportunity'. The opportunity came to the commercial animal health companies. By the turn of the century, many new products were made available to the poultry industry, many accompanied by claims of their beneficial effects on necrotic enteritis. The author's personal experience was that products based upon mannan oligosaccharides did support the bird's recovery but none of them replaced the benefits previously enjoyed when AGPs were used more freely.

3.3 Head Winds

The poultry meat industry continued to attract a level of negative publicity from some quarters. The problems discussed above generated the potential for a greater level of negative media interest. It is generally accepted by the industry that the extra focus applied by farming operations to overcome the problems experienced did deliver some benefit.

Government, non-governmental organizations and single-issue bodies all added pressure to an industry battling to deal effectively with the outcomes of removing AGPs. Around 2004, it was apparent that medication dealt only with the symptoms, but not the root causes driving necrotic enteritis.

A significant improvement occurred when farmers paid extra care to the details of the first two weeks of brooding. On the basis that the gut of the young bird is still developing during the first week of life, the thinking assumed sub-optimal temperatures may impact how their gut develops. As such, this could be a major contributor, if not the major cause, of failings of the gut system as the bird grows. As brooding techniques improved, the incidence of necrotic enteritis reduced. The industry was not without enteritis completely but housing conditions were improved by this change.

The learnings made at this juncture were, from a personal view, key to later successes achieved by the industry in reducing overall antibiotic usage. Some farmers started to realize improvements to disease control could be achieved by management changes on farms.

3.4 Antibiotic Stewardship

The BPC, which is the trade body representing the poultry meat industry in the UK, recognized the importance of protecting the efficiency of antibiotics and voluntarily formed its own Antibiotic Stewardship Scheme. The following is an excerpt from their website:

> BPC Antibiotic Stewardship ensures sustainable use of antibiotics; to protect the health and welfare of our birds, to safeguard the efficacy of antibiotics, and to produce food consumer's trust. As a result of BPC Antibiotic Stewardship, set up in 2011, the poultry meat sector became the first UK livestock sector to pioneer a data collection mechanism and share antibiotic usage data with the government's Veterinary Medicines Directorate (VMD). (BPC, 2017)

As the founding chair of this Stewardship scheme, it was clear that the industry needed to identify the size of the problem being tackled and put in place actions to monitor and trend usage over time. Concurrently, the industry was required to work collaboratively and share best practice in order that downward trends in usage could be demonstrated. The Stewardship therefore agreed the following actions:

1. To quantify accurate usage data. The only usage data widely available at that time were based on UK sales of antibiotics. These data lacked sufficient granularity to identify which species of livestock had received the product or in what timeframe the product was administered. No meaningful analysis outside of gross sales by product type was available. Greater clarity was required by the industry.

2. To monitor and reduce overall usage over time. Stewardship members agreed to provide the BPC with usage of antibiotics both by type and by quantity of active ingredient of each product. Providing the liveweight produced (in kg) allowed a comparative metric of milligrams of antibiotic per kilogram of production. The first year of data collection highlighted an overall lower usage of antibiotic by BPC members than inferred through sales data. Ongoing collection and trending of these data allowed the Stewardship to set actions in place to achieve reductions in both quantity and/or antibiotic type.

3. To reduce the critically important antibiotics as described by the WHO (WHO, 2011). The list of 'critically important antibiotics,' later revised in 2013 to HPCIAs, included antibiotics from the following classes: quinolones, third- and fourth-generation cephalosporins, macrolides, ketolides and glycopeptides. In most cases, for poultry, alternative efficacious antibiotics were available that allowed the Stewardship to significantly reduce HPCIAs. In January 2012, the Stewardship agreed to stop using third- and fourth-generation cephalosporins. A short while later, it was agreed to apply further restrictions on the use of quinolones. It should be noted, in restricting usage it was accepted quinolones could be used under exceptional circumstances. Notably, where there has been a full veterinary investigation and clinical diagnosis, no alternative antibiotic is available and usage is supported at executive board level. The usage of macrolides was kept under review, the aim being to lower usage where possible. The unique role macrolides had in the treatment of certain specific poultry diseases, e.g. mycoplasma, was, however, recognized by the Stewardship. After the ban on AGPs, glycopeptide antibiotics were not used in the poultry industry.

4. To stop prophylactic usage of antibiotics. For clarity, the following definition of prophylactic treatment will be used and is that adopted by Responsible Usage of Medicines in Agriculture Alliance (RUMA): 'Treatment of an animal or a group of animals, before clinical signs of infectious disease, in order to prevent the occurrence of disease or infection' (RUMA, 2013).

At the start of the Stewardship, it was clear the administration of antibiotics prophylactically was open to criticism. Fundamentally, treating animals before clinical signs of infectious disease were manifested could frequently result in fit, healthy animals being treated. It is the author's opinion that in 2011 most of the day-old broiler chicks were treated on arrival at the farm (day-old medication). Treatment was applied prior to those actual chicks being diagnosed with disease. This situation needed addressing. Increasingly, Stewardship members adopted strategies that eventually resulted in most chicks not receiving day-old medication.

5. To share information between Stewardship members. The collaborative spirit extant within the Stewardship encouraged members to share successes and failures of strategies to improve husbandry, hygiene and management practices. The post-adoption impact of such strategies was then reviewed against the need to use antibiotics. Those strategies found to be a success could be adopted by others, while the failures could be avoided. The collective action of all allowed quicker advances to be made over time. It should be remembered that the Stewardship was a voluntary collaboration by industry competitors to achieve a common goal. It was imperative, if success was to be achieved, that the process must not become competitive. Sharing ideas, celebrating success and supporting each other was crucial. It was not the speed of travel of individual members along what was effectively a journey of improvement that was important. The important issue was that each member improved and could demonstrate falling trends in antibiotic usage. The industry voluntarily taking this course of action had the luxury of time on its side. Legislative and regulatory bodies were supportive and allowed the industry to progress unencumbered by external stricture. This collaborative approach has recently been published as a case study by the Food and Agriculture Organization of the United Nations (FAO) (FAO, 2022).

6. To communicate and share progress publicly. One of the real strengths in the BPC Stewardship was its willingness to share the results of its work publicly. In 2016, the first of the BPC's Antibiotic Stewardship Scheme reports was released and subsequent reports have been released on a more or less annual basis (BPC, 2016). The sharing process included invitations to external bodies such as the Veterinary Medicines Directorate (VMD) to attend the working meetings.

3.5 Making the Difference at Farm Level

In short, an Antibiotic Stewardship Scheme can only be effective if those who use the antibiotics, in this case the poultry farmers, take decisive actions. The greatest initial challenge faced was encouraging a change of culture across the farming community, or so we thought. It is important to appreciate that the members of the BPC Stewardship were, in the main, employees of large integrated companies or veterinarians. A significant percentage of the supplying farms to those companies were private farms who held supply contracts to the larger integrated businesses. To make meaningful changes to the usage of antibiotics it was essential that all farms, both corporate and privately owned, were willing participants in the process.

At farm level, with the support of attending veterinarians, pressure from retail customers and changes in standard requirements of farm accreditation schemes, overcoming inertia to change was a surprisingly smooth process. There were, of course, those farmers less responsive to innovative ideas. That was, however, to be expected and just needed industry players to coach and help such individuals. Mindful from the start of the process that farms would not all move along the journey at the same speed, it was essential, one way or another, to win them over to the idea that the way the industry operated at that time was not an acceptable practice going forward.

To win over the wider farming community amongst the Stewardship members, the messages shared included the following:

- *Today's practice of using antibiotics is not acceptable practice for the future.* Once this point was accepted, moving forwards was inevitable.
- *If the industry does not initiate and control antibiotic usage it was highly likely that Government would take steps to do so.* Industry experience is that it is better to bring about change through agreed industry action than be subject to blunt Government regulation.
- *Through self-regulation, the industry can bring about change through evolution over time.* This approach would be better than following a government-initiated change agenda, which may be set to a more aggressive timescale than the industry would feel comfortable achieving.

3.6 Focusing on the Detail of Bird Husbandry

As individual farms accepted the need to reduce antibiotic usage, the question then was how to achieve it. The use of antibiotics in the face of bacterial disease is justified following veterinary investigation, diagnosis and potentially sensitivity testing. Experience showed, however, that too frequent antibiotic usage masked sub-optimal farming practice. Realistically, farms discovered the weakness in their husbandry, hygiene and management of livestock when antibiotics were removed from the equation. The phrase 'a farmer finds out how bad their farm management of livestock is when antibiotics are removed for the first time' was sadly true.

No single factor is universally at fault across a group of farms leading to the need for antibiotic treatment. Each farm has a distinct set of factors that collectively can create problems. The list below, however, highlights those facets of farm management which the authors believe are commonly contributory factors. As such, these are the ones that it would be sensible to investigate as a priority in any livestock operation attempting to reduce usage.

- *Poor house hygiene post cleanout*: achieved by microbiological swabbing post cleaning and disinfection, to include swabbing of floors, walls, drinkers, feeders and inlets.
- *Poor drinker line sanitation*: achieved by shock sanitation of drinker lines during hygiene phase to remove biofilms.
- *Poor water sanitation*: continual low-level dosing of appropriate water sanitant throughout the birds' lives.
- *Lacking water hygiene assessment*: microbiological testing of water cleanliness to prove efficacy of sanitization.
- *Poor staff hygiene*: staff must have house-dedicated footwear/clothing ideally operated through a barrier system on entrance to each house.
- *Incorrect bedding hygiene and depth*: the bedding must be clean and dry, avoid bedding that has been poorly stored in damp conditions. The bedding must be of a depth to insulate the chicks from the concrete floors. Mouldy bedding must be avoided at all costs and the bedding must easily cover all of the concrete floor.
- *Incorrect or variable brooding temperature across the entire floor area*: frequently the temperature recorded at the sensor may be correct. That does not mean, however, that temperature is even across the whole house. Young chicks can become chilled near the walls of the house or too hot near the heaters.
- *Insufficient ventilation*: a minimum ventilation plan must be operating, matching the need for gaseous exchange and moisture removal with the biomass contained within the building. Additional ventilation must be available to then contain house temperatures within planned set points. A further aid to the industry was the implementation of renewable heat incentives (RHIs) to encourage less use of fossil fuels. This innovation introduced the concept of indirect heating of poultry houses, which reduced the moisture level in houses and thus made ventilation and house environmental management easier.

- *Poor management training*: the single most important influence on the health and welfare of farm livestock is the skill of the personnel charged with looking after them daily. A proficient level of knowledge and expertise in health, welfare and hygienic practices is essential. Training must be planned, recorded and refreshed; it cannot be an ad hoc event conducted when time allows.
- *Poor observation skills of farm management*: staff must be encouraged/trained to use all their senses when operating in and around livestock. Too frequently, basic problems can go unnoticed by staff who do not bother to look around as they work. A plastic office chair is a surprisingly effective way of developing farm skills. Such a chair should be placed within the bird area of each house. The stockman should be instructed to sit and observe bird activity and use his or her nose to check the air for freshness. A period of 5 minutes carrying out these two tasks before doing anything else in that house will pay dividends to the farm. In the initial stages, to help the staff develop their skills, the individual should describe what they observed and what appropriate action they might take.
- *Failing to monitor and record performance and environmental parameters*: farms that can present good, detailed records and can demonstrate comparison of performance and environmental parameters achieved against standards will facilitate investigations during the onset of disease.

3.7 Changing the Mindset to Disease Management

Measuring and checking alignment across the 11 areas listed above will aid a farmer in reducing insults to the bird and thereby their need for veterinary intervention and treatment.

Historically, if birds become sick, the accepted process is that a veterinarian conducts an investigation, gives a clinical diagnosis and may, depending upon the outcomes, recommend treatment with antibiotics. The treatment is administered and hopefully the birds recover to full health. It is at this point that the farmer and the vet should hold an open and full review. The aim of this review was to understand what changes need to be made to avoid a similar issue occurring in the future. It is common that the accompanying veterinary report submitted following investigation, besides recommending treatment, will have proposed a number of recommendations. These recommendations frequently require changes in management processes.

The desire of humans to seek out answers seems to be second nature in most situations. In a car crash, the aim is to understand what caused the crash and who is to blame. In a house fire, once the fire is out the first question is 'how did the fire start?', presumably so that mitigation can be put in place to avoid reoccurrence. In terms of animal disease, historically once the animals have responded and are well that same inquisitive instinct fails in many cases. To effectively reduce antibiotic usage over the long term, the post-disease review is crucial. This review must identify every single aspect that needs attention to prevent reoccurrence. Unfortunately, experience shows that even constructive criticism of how we manage tends not to be sufficiently accepted to a point where required actions are left undone. Those farmers adept at challenging their own operational standards and responding to all constructive criticism or advice proved to be the ones most capable of reducing antibiotic usage over the long term, with the reverse also true.

In other words, to be successful in farming poultry with little or no antibiotic usage, farm personnel need to accept that shortfalls in their own management can be a significant risk factor to be considered. Critique of one's own management is, in the author's view, a significant factor in the success, or otherwise, of preventing intestinal disruption and subsequent infection spreading within a group of birds.

3.8 Managing the Microbiome

Understanding of the crucial role the microbiome plays in the health and vitality of our livestock allowed more elegant strategies to be developed in supporting the microbiome and thereby the health of the birds. Some farmers started to explore methods aimed at encouraging the development of the bacterial populations throughout the gastrointestinal tract.

Dr Stephen Collet of the University of Georgia first used the phrase 'Seed, Feed and Weed' when applied to poultry gut health (Collet, 2012). The objective of this process is to encourage early onset of a biodiverse bacterial gut population that restricts the opportunity for pathogenic bacteria to colonize the same space. The three steps in the process he proposed are outlined below:

1. *Seed*: the newly hatched chick gut is naïve; it lacks a developed microbiota. In this step, the chicks are sprayed or fed bacteria to start the process of seeding their gut with preferred types of bacteria. It could be a single species such as a *Lactobacillus* or a mix of bacteria as provided in a number of proprietary competitive exclusion products. The aim is to start colonization with 'good bacteria'. The earlier this colonization started, the quicker a mature microbiota is likely to develop and the options for pathogenic bacteria colonization diminish. Speed is of the essence; seeding is normally achieved by either spraying day-old chicks at the hatchery, feeding treated feed or providing bacteria via the drinking water.

2. *Feed*: in this phase, the aim is to encourage growth of the newly seeded bacteria by adding appropriate nutrients. For example, the addition of butyrate to the feed provides butyric acid in the gut, an acid upon which many bacteria seem to thrive. It is important to note, however, that the feeding phase will also support the development of many pathogenic bacteria.

3. *Weed*: the weeding phase is crucial; its aim is to remove many types of pathogenic bacteria. This can be achieved in a number of ways; commonly the addition of a type 1 fimbriae blocker is used and an example of such a blocker is a commercially available product called Actigen®. Pathogenic *Salmonella* and *Escherichia coli* attach to the cell wall via filaments called fimbriae. The type 1 fimbriae blockers agglutinate the bacteria and the blocker, along with the agglutinated bacteria, pass down the gut lumen and are expelled with the faeces. This phase effectively limits the number of pathogenic bacteria competing for a niche in the gastrointestinal tract.

The aim of these three steps is to increase diversity of the gut bacterial populations. To be successful this process requires farmers to be committed to moving away from antibiotic usage. The indiscriminate impact of antibiotics on many types of bacteria can be deleterious to the objective of a Seed, Feed and Weed programme. Usage of antibiotics on farms using a Seed, Feed and Weed programme should be avoided wherever possible. The author's experience is that a Seed, Feed and Weed programme, carried out with diligence and care, significantly enhances the ability of the farmer to grow birds with very little need for antibiotic intervention.

3.9 Conclusion

The experience of the UK poultry meat industry demonstrates that it is quite possible to significantly reduce the usage of antibiotics within poultry production through the adoption of an Antibiotic Stewardship Scheme. The industry managed to reduce usage in chicken production by 85% over a four-year period; similar reductions were recorded in turkey production. This reduction was achieved by a process which included the following six steps:

1. *Independent Governance*: active industry collaboration under the BPC Antibiotic Stewardship Scheme, which offered leadership/ governance and anonymity.
2. *Quantifying the problem*: through collection and compilation of accurate usage data.
3. *Ownership of the problem*: early acceptance by the executive and senior management of participating companies that the current antibiotic usage levels were not acceptable.
4. *Communications*: regular feedback to industry on trended antibiotic usage by species and by type; sharing of successful reduction strategies between participants.
5. *Selection and prioritization of goals*: reducing and/or eliminating usage of highest priority critically important antibiotics; reducing and/or eliminating prophylactic administration of antibiotics; understanding drivers for the most frequently used antibiotic classes.
6. *Sharing information publicly*: agreement to publish results and trends over time demonstrates a level of commitment and integrity to the process.

The above six-step process was adopted successfully; and this success is a credit to the participating industry members who diligently worked to achieve a collective benefit for the industry. The author accepts that the integrated nature of the UK poultry meat industry probably made the process easier to implement than in other less integrated livestock sectors. Integrated industries tend to have fewer decision makers throughout so decisions can, potentially, be made quickly. The adoption of these six steps could be applied to any livestock industry interested in achieving a reduction in antibiotic usage. Industries with a less integrated structure will find the process potentially more challenging but entirely possible.

It is the author's personal view that to be successful in reducing antibiotic usage in livestock production, producers must do the following five things well:

1. Those involved in farming stock must accept and believe that producing livestock without antibiotics is necessary and achievable. In absence of this belief the process will falter.
2. Those involved must be prepared to challenge every facet of their production process. This challenge can only realistically be achieved through measurement and comparison of results to industry best practice. If a producer can start with an honest critical evaluation of his or her farming operation and practices and is prepared to take on alternative ways of operating, success will follow.
3. Antibiotics are a treatment of last resort only.
4. Focus on livestock keepers. Antibiotics are administered to livestock usually by farm staff so it is imperative to work hard to educate, train and help the livestock keeper.
5. Celebrate success, do not berate failure.

A final observation, as a member of the UK poultry meat industry, is that it was notably evident in chicken and turkey production that the fewer times antibiotics were used, the more robust the birds became. How could this be the case? It is reasonable to surmise that broad-spectrum antibiotics may solve one problem but create collateral damage to the very bacteria the Seed, Feed and Weed process is there to promote.

Acknowledgements

I would like to thank the UK poultry meat industry members for their diligent work, work which resulted in such a dramatic reduction in antibiotic usage. On behalf of the Stewardship, I thank Slate Hall Veterinary Practice, who designed collection templates and performed the bulk of the collation and analysis of data from individual members. And last but not least, I would like to thank the Executive and staff of the British Poultry Council and notably Maire Burnett who worked tirelessly keeping us all on track, providing us with up-to-date information and ensuring data remained anonymous.

References

BPC (2016) Antibiotic stewardship scheme. Available at: http://www.farmantibiotics.org/wp-content/uploads/2016/11/The_BPC_Antibiotic_Stewardship_Scheme_April2016.pdf (accessed 23 January 2023).

BPC (2017) Antibiotic stewardship report June 2017. Available at: https://britishpoultry.org.uk/download/bpc-antibiotic-stewardship-report-2017/ (accessed 25 January 2023).

Casewell, M., Friis, C., Marco, E., McMullin, P. and Phillips, I. (2003) The European ban on growth-promoting antibiotics and emerging consequences for human and animal health. *Journal of Antimicrobial Chemotherapy* 52, 159–161.

Collet, S. (2012) *The "Seed, Feed and Weed" approach to managing intestinal health in drug free production systems.* In: *The Poultry Informed Professional*, University of Georgia, Georgia. Available at: https://www.researchgate.net/publication/303443576_The_role_of_feed_composition_on_gut_health_in_poultry (accessed 25 January 2023).

FAO (2015) Codex texts on Foodborne antimicrobial resistance. Available at: https://agris.fao.org/agris-search/search.do?recordID=XF2016078930 (accessed 23 January 2023).

FAO (2022) Tackling antimicrobial use and resistance in food-producing animals. Available at: https://www.fao.org/family-farming/detail/en/c/1619167/ (accessed 25 January 2023).

Kirchhelle, C. (2018) Pharming animals: a global history of antibiotics in food production (1935–2017). *Palgrave Communications* 4, 96.

RUMA (2013) Preventative use statement. Available at: https://ruma.org.uk/wp-content/uploads/2022/07/The-Preventive-Use-of-Antibiotics-in-Farm-Animals-Prophylaxis.pdf (accessed 25 January 2023).

WHO (2011) World Health Organisation: Critically important Antimicrobials for human medicine, 3rd revision. Available at: https://www.who.int/publications/i/item/9789241504485 (accessed 23 January 2023).

WHO (2016) World Health Organisation: Critically Important Antimicrobials for Human Medicine 5th revision. Available at: https://www.who.int/publications/i/item/9789241512220 (accessed 23 January 2023).

CHAPTER 4
Gut Health in Commercial Feed Formulation

Ralph Bishop[1] and Dawn Scholey[2*]

[1]*Premier Nutrition, Rugeley, UK;* [2]*Nottingham Trent University, UK*

Abstract

This chapter will cover the general practice of poultry gut health maintenance from a poultry nutritionist standpoint. Gut health begins at the hatchery and it is important to maintain good husbandry and biosecurity from this point onwards. There are several strategies to promote good gut health in the early post-hatch period, including hatching chicks on farm and ensuring early access to feed and water. Once the chicks have started to eat and grow, strategies such as managing diet formulation and precision feeding can be used to maintain bird health. Nutritionists need to consider the raw materials they are using when formulating and use their knowledge of these to add appropriate enzymes and additives to support bird health. Practitioners need to be taking an holistic view of gut health to maintain optimal performance.

4.1 Introduction

There has been substantial coverage in the mainstream media around human gut health, particularly in relationship to food intake. One recent report highlighted the important of provision of prebiotics and the role of food intake in the incidence of leaky gut (Callingham, 2022). These issues are also prevalent in poultry production, with the rapid growth and high-performance requirements of the sector enhancing the importance of the optimal gut health of birds commercially. Commercial formulators and nutritionists need to be considering the gut health of the bird all the time, as general good practice. Ensuring diets and husbandry are managed as well as possible is vital to maintaining a healthy

*Corresponding author: dawn.scholey@ntu.ac.uk

DOI: 10.1079/9781800622746.0004

gut and hence a healthy and productive flock. In this chapter, general practice measures for ensuring gut health will be discussed, along with a formulators perspective on managing diets for gut health. The commonly used additives for managing flock health will also be discussed.

4.2 Strategies for Improving Gut Health in Young Birds

Gut health begins at the hatchery and requires good husbandry and biosecurity combined with best practice in terms of care of young chicks. There are several procedures that can help the growth and development of the gastrointestinal tract in the newly-hatched chick, and this is vital for optimal nutrient digestion. Eggs are traditionally hatched in a specialized facility and then transported within the first few days to the production site. However, this causes stress to the chicks and does not allow access to feed and water. The industry is considering alternatives to this hatchery system, including on-farm hatching and early access to feed at the traditional hatchery.

4.2.1 On-farm hatching

Traditionally, eggs are hatched in a designated hatchery and then transported within the first 72 hours to the farms for rearing. However, there is potential for eggs to be hatched on the rearing facilities, which would allow the birds to have immediate access to feed, water and light, and would reduce handling stress. On-farm hatching of eggs has been compared with traditionally hatched birds, and although chick quality was worse at day one, the farm hatched chicks were heavier and maintained this until day 21 (de Jong *et al.*, 2020). The same study also found decreased mortality for farm-hatched chicks and a reduction in litter moisture, leading to a tendency for fewer foot pad lesions. An earlier study also found improved week 1 performance and reduced prevalence of foot pad dermatitis compared to traditional hatching (de Jong *et al.*, 2019). This improvement may be due to a reduction in transport stress, as transportation of chicks has been shown to increase production of the stress hormone cortisol (Witjes *et al.*, 2022).

The downside of this system is that farms would need designated facilities for hatching, with good biosecurity and staff training. This obviously comes with a cost that may not be covered by improvements in bird growth as it is unclear whether improvements in weight would be maintained to slaughter.

4.2.2 Early feeding

Early feeding is the process of ensuring the chicks have access to feed, either a starter pellet or crumb, immediately post hatching. Chicks are precocial and will forage for feed as soon as they have hatched if it is available to them. Birds are traditionally hatched together over a 36-hour period, where they typically will not have access to feed and water. This results in bodyweight loss of 4 g per 24 h (Pinchasov and Noy, 1993). In a hatching chick, the yolk sac makes up around 15–20% of the bodyweight, with the chick using this for nutrition for the first 48 hours. However, if the birds have access to feed, this increases the rate of yolk utilization (Noy and Sklan, 1996). This may be due to the presence of the feed increasing the mechanical activity in the intestine. Early provision of nutrients also encourages the development of the gastrointestinal tract, which increases in size rapidly immediately post hatching. The absorptive area of the small intestine is increased as crypts develop and villus heights are greater. It is clear that chicks that eat early, grow more. Appropriate nutrition as close to hatching as possible can improve uptake of nutrients, increase muscle growth and improve bird performance. It is claimed that this improvement will continue to marketing, with a reported 8–10% improvement in both bodyweight and breast yield (Noy and Sklan, 1998).

It may be beneficial for chicks to have post-hatch access to feed but hatching occurs over a longer period, which would mean that some chicks would have up to 36 hours more feed consumption than others. Also, variability of access to feed can be an issue. Both of these factors would potentially decrease flock uniformity, so the feed access needs to be managed to maximize both growth and consistency of the chicks.

4.3 Dietary Strategies for Improving Gut Health

Once the chicks are hatched and on the ground in the production facility, the emphasis moves from getting the chicks to eat, to focus on what they are consuming. It is important to formulate diets with a consideration for gut health throughout production, to maintain optimal performance, but also reduce the impact of pathogens. In the next section, the impact of low-protein diets will be discussed and how to maintain gut health while minimizing protein use. The potential opportunities to improve nutrient supply via precision feeding methods will also be discussed.

4.3.1 Low-protein diets

There has been a drive towards reducing protein levels in order to minimize costs and mitigate the environmental impact of production, by reducing excreted nitrogen. However, this raises the question of whether low-protein rations make it harder to maintain gut health, due to increased fibre levels. In practice, generally it can be assumed that a certain inclusion of fibre is a positive thing for the gut development, but if the inclusion of barley, rye or oats is high then the formulator would look at the cereal enzyme use in the diet. For instance, rye is challenging for the birds but is of growing interest to the industry. It is important when using more challenging feed ingredients that they are used in appropriate quantities in the first instance. The nutritionist and producer then need to ensure they are using the right additives for the raw materials in the diet, and at the correct dose. If the product is appropriate for the raw material, then it may be appropriate to increase the dose with the higher inclusion of some cereals in order to mitigate issues.

It is important in low-protein diets to ensure the diets are as balanced as possible with any additives and enzymes as required by the situation and issues. Single amino acids are also commonly included, particularly lysine, methionine and threonine, with others less commonly used as standard practice, but this depends on the economics in a specific diet. Newer amino acids that are more commonly being added include valine, arginine and isoleucine with tryptophan also being conditionally essential (for example in soya-free diets). The use of added amino acids is becoming economically attractive as soya price can be volatile. Feeding adequate levels of limiting amino acids increases flock uniformity (Berhe and Gous, 2008) with the potential to reduce variation in bodyweight by up to 40% by adding sufficient methionine (Lemme, 2003). It is important to maintain sufficient amino acid supply to the diet, while considering that there will be bird-to-bird variation in requirement, dependent on size, sex and age.

The advantages of reducing crude protein are substantial and, as long as amino acids are adequate, it does not affect performance deleteriously (Cappelaere *et al.*, 2021). Lower protein consumption can minimize water use, as it has been shown that water intake is related to dietary protein in a dose-dependent manner (Huang *et al.*, 2011). This is turn helps manage litter moisture and reduces the occurrence of foot pad lesions (Van Harn *et al.*, 2019). Lower protein also leads to a lower litter nitrogen content as less is excreted (Belloir *et al.*, 2017), which helps minimize the environmental impact of poultry production.

Examples of how soya content can be lowered in diets with added amino acids are shown in Table 4.1 (for layers) and Table 4.2 (for broilers).

4.3.2 Precision feeding

Precision feeding is the process of meeting the requirement of the bird as accurately as possible through the different stages of production. Typically, three to five phases are used in practice, but the nutritional requirements of broilers alter daily (Aviagen, 2022), meaning that some nutrients are under or over supplied at difference phases of production. The first half of the production cycle has the greatest disparity between supply and requirement, and this could result in reduced growth and increased costs (Moss *et al.*, 2021). A reduction in nutrient retention can also encourage the growth of pathogens (Adams, 2004).

The number of phases fed could be increased to improve the precision of feed formulation for broilers. Daily feed manufacture as utilized in dairy farming is impractical for poultry but it may be possible to precision-blend different feeds on farm to produce a similar effect. Precision feeding based on feed blending has been utilized in pigs, resulting in a reduction of the environmental impact of production owing to the reduction in excretion of excess nutrients (Andretta *et al.*, 2014). Hauschild *et al.* (2015) attempted this type of

Table 4.1. Example layer diet with added isoleucine.

Raw material	Control	Plus isoleucine	Difference
Wheat	48.976	51.513	2.537
Barley	15.000	15.000	
Hipro sunflower extracted	9.208	9.475	
Brazilian soya meal	13.652	10.844	–2.808
Soya oil	2.829	2.581	–0.248
Limestone	9.133	9.11	
Salt	0.205	0.180	
Sodium bicarbonate	0.263	0.302	
Monocalcium phosphate	0.159	0.178	
Lysine HCl	0.155	0.236	
DL Methionine	0.159	0.178	
Threonine	0.012	0.048	
Isoleucine	0	0.050	
Valine	0	0.024	
Premix	0.250	0.250	
PROTEIN (%)	16.93	16.11	–0.818
OIL B (%)	4.88	4.63	–0.252
LINOLEIC (%)	2.36	2.23	–0.129

Table 4.2. Example broiler grower diet with added arginine.

Raw material	Control	Plus arginine	Difference
Wheat	62.709	64.259	1.550
Whole oil seed rape	5.000	5.000	
Brazilian soya meal	27.284	25.713	–1.571
Soya oil	1.911	1.742	–0.169
Limestone	1.118	1.123	
Salt	0.158	0.144	
Sodium bicarbonate	0.246	0.268	
Monocalcium phosphate	0.546	0.558	
Lysine HCl	0.293	0.340	
DL Methionine	0.317	0.329	
Threonine	0.136	0.158	
Arginine	0	0.044	
Isoleucine	0	0.014	
Valine	0.032	0.058	
Premix	0.250	0.250	
PROTEIN (%)	21.31	20.91	–0.405
OIL B (%)	6.16	5.99	–0.172
LINOLEIC (%)	2.30	2.21	

programme, blending two feeds to create 14 different phases over the 42-day production cycle. These authors found an increase in daily weight gain and also in yield. Sharma *et al.* (2014) tried a simpler method of feed blending, using a starter diet blending with whole wheat to create ten phases, but found that this compromised feed conversion ratio of the birds because the nutrient formulation of these phases was not well balanced. A more effective method may be to blend a low-protein, energy-dense concentrate with a protein-dense concentrate on a daily basis, as required. Both these diets could be pelleted to maximize feed intake. This procedure has been investigated using two pelleted diets blended daily to create an optimal energy protein diet for the birds' age and stage (Moss *et al.*, 2020). These authors found no significant differences in performance but did report an increase in the dressing percentage. They also reported an increase in sales revenue of the equivalent of £0.78 (GBP) per bird ($1.30 AUD) owing to the increased meat yield. This suggests that the technology required for this precision feed blending may be able to pay for itself in short order if these types of yield increases are typical.

Precision feeding could also have value in feeding broiler breeders, where feed restriction is required to maintain lay and achieve high productivity. An individual feeding procedure was utilized to manage feed intake across a flock of broiler breeders where uniformity is typically poor due to competition for limited feed supply. By individually feeding birds according to their weight, the uniformity of the flock was reduced to less than 2% by week 20 (Zuidhof *et al.*, 2017). Although this type of individual feeding is prohibitive from a practical and cost perspective, there may be uses for precision feed blending in broiler breeders to manage weight gain.

4.4 Raw Materials

The type of raw material incorporated in poultry diets will depend on several factors. The availability of the raw material in the required geographical area, combined with cost, will define diet inclusion but the quality and variability of the material also needs to be considered. Many feed ingredients contain elements that make them less digestible or that have adverse effects on the bird. These anti-nutritional factors (ANFs) are described below.

4.4.1 Anti-nutritional factors

Formulators must consider ANFs when producing diets, dependent on the raw materials incorporated. ANFs include tannins, trypsin and protease inhibitors, lectins, glucosinolate and haemagglutinins in peas (Savage, 1989). These ANFs will typically be managed through diet formulations rather than testing as the latter is expensive and difficult. Usually, the dietary ingredients in question are added at levels that are considered manageable according to the literature; for example, peas and beans would typically be added at around 7–8%, and rarely added at levels higher than this unless economically driven.

In turkey and game diets, which are higher protein, it would be usual to bring in a second protein source to reduce the effect of ANFs, such as potato protein, gluten meal, sunflower meal or rapemeal. This helps keep trypsin inhibitors at a manageable level.

4.5 Additives: Enzymes

The use of additives in poultry diets must be based upon the overall value for the supply chain, with decisions being based on the costs and saving per kg of chicken produced and not based on cost per tonne of feed. However, some additives are cost prohibitive. Cereal enzymes and phytase are used as standard in poultry diets to decrease protein use and improve digestibility.

There are also other additions that are good practice but not typically included in the diet. These include butyrate and prebiotics, which can be helpful for good quality nutrition, but their addition will depend on the environment. For example, some situations make the use of these additives more appealing. If the site is older and has a high bacterial load, this provides more physiological challenge for the birds and this would suggest the use of a prebiotic additive to protect against the environmental challenge. The prebiotic would typically be used in starter diets where the risk to the birds is greatest but the feed intake is small, making the increased cost easier to justify.

Single-use probiotics are not commonly used in practice but tend to be used in situations such as bacterial challenges at hatching, rather than ubiquitously. The next section will cover the use of different feed enzymes, beginning with those included as standard and touching on the more infrequently used products.

4.5.1 Cereal enzymes

Plant cell walls contain non-starch polysaccharides (NSPs) dependent on the cereal type. Barley and oats contain β glucans, whereas wheat and rye contain xylans, both of which are complex carbohydrates with extensive cross-linking. They are water soluble and form highly viscous gels, which result in increased viscosity of digesta and hence reduced digestibility of nutrients. Xylans increase water intake in birds and this leads to high prevalence of sticky droppings and wet litter, with the associated increase in foot pad dermatitis (Khattak *et al.*, 2006), whereas β glucans also increase digesta viscosity, which affects protein and starch digestion (Hesselman and Aman, 1986). Poultry do not have the capacity to hydrolyse NSPs, so require the addition of exogenous enzymes to break the carbohydrate chains into smaller pieces. Cereal enzymes include xylanases, β glucanases and, less commonly used, amylases, and they can be supplied as individual enzymes or a mixture of several to deal with multi-cereal diets.

Cereal enzyme use is widespread in the industry and is cost effective. The appropriate enzyme depends on several factors, including the raw materials incorporated into the diet and the levels of deleterious carbohydrates present. Picking the best enzyme for the situation is key, as is applying it at a dose that has the greatest effect whilst still being economical. There are also interactions between different enzyme combinations, where the benefits are not always additive, and this needs to be taken into consideration.

4.5.2 Phytases

Phytase has been used in poultry diets since the 1990s and is now a standard addition throughout most of the world. Phytase is used to enhance the availability of phosphorus (P) from cereal phytate, by releasing the otherwise unavailable phosphorus bound to the inositol ring in cereal cell walls (Choct, 2007). This improvement in availability allows a reduction in the amount of inorganic phosphorus in diets, which is more cost effective as inorganic P is limited in quantity and hence can be costly. The price of feed phosphate has increased considerably in the years up to 2022, making its heavy use in diets prohibitively expensive. Ensuring the cereal P is digested has the knock-on effect of reducing excretion of undigested P, which helps mitigate the environmental impact of poultry production. Phytate also has other extra phosphoric detrimental effects on mineral, energy and amino acid metabolism, which addition of phytase can alleviate (Cowieson *et al.*, 2011).

Typically, phytase doses were added at 500 FTU/kg but over the last 15 years it has become more common to add high, super-doses of phytase to diets. High doses of up to 10,000 FTU/kg have been investigated and show additional benefits (Pirgozliev *et al.*, 2007). Super-doses of phytase confer substantial value, probably due to a combination of factors. These include increased P release, reduction in phytate esters and hence a reduction in their antinutritive effects, and the generation of myo-inositol, which may itself have beneficial effects.

4.5.3 Proteases

Proteases typically work best in the starter phase, with a reduced efficacy from grower phased onwards. As protease affects nutrient utilization and hence growth, it may be that the young undeveloped gut does not produce sufficient digestive enzymes, making an exogenous supplement useful. Initial work on proteases was carried out more than 60 years ago, with some authors reporting no effect (Jensen *et al.*, 1957) and some reporting an improvement in performance (Burnett, 1966). It may be that supplementation of protease reduces the requirement for exogenous amino acids and it has been reported that amino acid utilization is improved in diets high in ANFs such as trypsin inhibitors (Ghazi *et al.*, 2002). It has been demonstrated *in vitro* that protease could deactivate trypsin inhibitors in just over an hour (Huo *et al.*, 1993). In laying hens, it has been suggested that crude protein could be reduced by 1% in diets with protease supplementation (Yadav and Sah, 2006).

4.5.4 Other enzymes

There are other enzymes that are less commonly used in poultry production, such as lipases to enhance lipid digestion and galactosidases to help neutralize ANFs in legumes. However, they can have limited advantages and cost of addition is therefore uneconomic. Lipase, amylase and amyloglucosidase were found to have no benefits on any parameters in young chicks (Slominski *et al.*, 2004) so the evidence for value of addition is limited.

4.6 Other Additives

4.6.1 Butyrate

Butyrate is the additive most frequently added into diets in practice, particularly in starter diets where the risk to the birds is greater and the cost is easier to justify. *In vivo*, butyrate can be produced in the large intestine of the bird via bacterial fermentation of undigested carbohydrates (Tan *et al.*, 2014). Generally, butyrate has a positive effect on performance, which is mainly evident in younger birds (Bedford and Gong, 2018). In broilers, butyric acid added to the diet can improve gut health and ensure the gut microbiome is balanced (Yang *et al.*, 2018). Intestinal integrity and immunity are also improved (Zou *et al.*, 2019). In addition, when used as a feed additive, butyrate can protect against pathogenic bacterial infections such as *Salmonella*. *Salmonella* colonizes the caeca, so any added butyrate needs to be available in this part of the gastrointestinal tract; it is therefore important to use formulations that increase caecal concentrations (Onrust *et al.*, 2020).

4.6.2 Coccidiostats

Coccidiosis costs US$3 billion globally, both in terms of treatment costs and loss of production. It is standard practice to use a coccidiostat, in combination with chemicals and monitoring, and to rotate to maximize efficacy. However, traditional products can increase the prevalence of resistant strains. There is potential for use of natural coccidiostats, such as phytobiotic-based products, particularly as standard products can have a negative consumer image because they tend to be associated with antibiotic use These tend to be saponin based, which is a glycoside that may act as an anticoccidial and immunostimulant. Saponin has been incorporated in vaccines to enhance immune response (Xiao *et al.*, 2009) and also has been suggested to improve gut architecture (Cheeke, 2009), but the evidence is limited for the benefits at present. These are of most interest in the pheasant and turkey markets, where they would be targeted at the most challenging points of rearing. In organic production, coccidiostats are banned and, in this case, vaccination can be effective, but also increases cost of production and this can be prohibitive. Enzymes may also have positive effects on coccidiosis (Morgan and Bedford, 1995).

4.6.3 Oligosaccharides

There is a wide variety of compounds that are proposed to act as a prebiotic, including inulin and a variety of oligosaccharides, such as xylo-oligosaccharides

(XOS), mannan-oligosaccharides (MOS) and fructo-oligosaccharides (FOS). Some of the more commonly used products will be detailed below. It is worth noting that prebiotics tend to be of most benefit in challenging environments, such as old sheds or where the biosecurity is poor. They seem to be of less benefit when used in clean, well-managed facilities.

In wheat diets, the application of xylanase cuts the arabinoxylans into shorter chains of xylo-oligosaccharides. These shorter xylan chains are suggested to act as a prebiotic and stimulate an immune response, both when added as a supplement and when created *in vivo* by xylanase (De Maesschalck *et al.*, 2015). XOS have been shown to modulate the microbiota with an associated improvement in bird performance, but care must be taken when choosing the most appropriate dose (Ribeiro *et al.*, 2018). For instance, one recent study found no effect on bird weight gain (Singh *et al.*, 2021), whereas another study with a tenfold increase in dose did find a performance benefit (Zhenping *et al.*, 2013). Recently, studies have also found benefits of feeding XOS in laying hens (Zhou *et al.*, 2021), where improvements were seen in nutrient utilization and small intestinal architecture.

MOS are derived from yeast cell walls as a by-product of the brewing or baking industries. There has been a substantial weight of literature examining the use of MOS as a feed supplement, which has shown improved villus height (Iji *et al.*, 2001) and reduction in pathogenic load (Castillo *et al.*, 2008).

FOS are derived from plants via enzymatic synthesis. There have been improvements seen in broiler performance with FOS supplementation, with authors reporting heightened immune responses (Al-Surragai and Al-Khalaifah, 2022), improved mucosal structure (Xu *et al.*, 2003) and that FOS were able to positively alter the microbiome (Shang *et al.*, 2018).

Inulin is a carbohydrate supplement produced from plants, typically artichoke or chicory. The effects on bird performance with inulin supplementation are mixed (Buclaw, 2017) and it may only be effective in challenging environs. There are suggestions that inulin is more effective when supplemented early and it is important to pay attention to the dosage as increased dose often results in less efficacy (Velasco *et al.*, 2010).

4.6.4 Essential oils

Essential oils (EOs) are derived from plants as an oily mix of metabolites, usually with a low boiling point. There are a large number available, many of which are reported to have antimicrobial properties and module the gut microbiome (Rota *et al.*, 2004). Some oils such as cinnamon and oregano are suggested to be antioxidants with potential positive effects on meat quality. The strong scent of many EOs can improve feed intake due to stimulation of digestion but evidence of this palatability benefit is mixed (Brenes and Roura, 2010).

Variability in products can be high owing to differences in plants grown in various geographical areas, and also differences in growth conditions and harvesting. Windisch *et al.* (2008) summarized some EO studies and concluded that feed intake was reduced with no effect on weight gain, but many other authors have found no effects (Jang *et al.*, 2007). Currently the evidence for the efficacy of many EO products is limited.

4.7 Conclusion

To maintain gut health for poultry through the production cycle while maximizing output, there needs to be a combination of strategies. Stockmanship is important because there needs to be continuous feedback and monitoring of bird health and performance to allow issues to be mitigated as soon as they occur. Diet formulations also need to take into account raw material availability, price and, if possible, quality. It is not always possible to monitor all raw material batches for quality, nutritional content and the presence of negative factors but, where this can be done, this information can be used to put in place appropriate mitigation. It is important to be consistent in terms of application and the use of additives and, where possible, keep the formulation and process as simple as possible. This allows simple changes to be made quickly and efficiently.

References

Adams, C.A. (2004) Nutricines in poultry production: focus on bioactive feed ingredients. *Nutrition Abstracts and Reviews Series B: Livestock Feeds and Feeding,* CABI, Wallingford, UK, pp. 1N–12N.

Al-Surrayai, T. and Al-Khalaifah, H. (2022) Dietary supplementation of fructooligosaccharides enhanced antioxidant activity and cellular immune response in broiler chickens. *Frontiers in Veterinary Science* 9, 857254.

Andretta, I., Pomar, C., Rivest, J., Pomar, J., Lovetto, P.A. and Radunz Neto, J. (2014) The impact of feeding growing-finishing pigs with daily tailored diets using precision feeding techniques on animal performance, nutrient utilization, and body and carcass composition. *Journal of Animal Science* 92, 3925–3936.

Aviagen (2022) Ross 308 performance objectives. Available at: https://eu.aviagen.com/assets/Tech_Center/Ross_Broiler/RossxRoss308-BroilerPerformanceObjectives2022-EN.pdf (accessed 23 March 2023).

Berhe, E.T. and Gous, R.M. (2008) Effect of dietary protein content on growth, uniformity and mortality of two commercial broiler strains. *South African Journal of Animal Science* 38, 293–302.

Bedford, A. and Gong, J. (2018) Implications of butyrate and its derivatives for gut health and animal production. *Animal Nutrition* 4, 151–159.

Belloir, P., Méda, B., Lambert, W., Corrent, E., Juin, H. *et al.* (2017) Reducing the CP content in broiler feeds: impact on animal performance, meat quality and nitrogen utilization. *Animal* 11, 1881–1889.

Brenes, A. and Roura, E. (2010) Essential oils in poultry nutrition: Main effects and modes of action, *Animal Feed Science and Technology* 158, 1–14.

Buclaw, M. (2017) Inulin in poultry production. *World's Poultry Science Journal* 73, 301–308.

Burnett, G.S. (1966) The effect of damaged starch, amylolytic enzymes, and proteolytic enzymes on the utilization of cereals by chickens. *British Poultry Science* 3, 89–103.

Callingham, F. (2022) Leaky gut could allow toxins into the bloodstream. Available at: https://www.express.co.uk/life-style/health/1656668/how-to-live-longer-prebiotic-fibre-prevent-leaky-gut (accessed 29 March 2023).

Cappelaere, L., Le Cour Grandmaison, J., Martin, N. and Lambert, W. (2021) Amino acid supplementation to reduce environmental impacts of broiler and pig production: a review. *Frontiers in Veterinary Science* 8, 689259.

Castillo, M., Martin-Orue, S., Taylor-Pickard, J., Perez, J. and Gasa, J. (2008) Use of mannan-oligosaccharides and zinc chelate as growth promoters and diarrhoea preventative in weaning pigs: Effects on microbiota and gut function. *Journal of Animal Science* 86, 94–101.

Cheeke, P.R. (2009) Application of saponin as feed additives in poultry production. In: *Proceedings of 20th Annual Australian Poultry Science Symposium*, Sydney, Australia, pp. 50–55.

Choct, M. (2007) Enzymes for the feed industry: past, present and future. *World's Poultry Science Journal* 62, 5–16.

Cowieson, A., Wilcock, P. and Bedford, M. (2011) Super-dosing effects of phytase in poultry and other monogastrics. *World's Poultry Science Journal* 67, 225–236.

de Jong, I.C., Gunnink, H., van Hattum, T., van Riel, J.W., Raaijmakers, M.M.P. *et al.* (2019) Comparison of performance, health and welfare aspects between commercially housed hatchery-hatched and on-farm hatched broiler flocks. *Animal* 13, 1269–1277. doi: 10.1017/S1751731118002872.

de Jong, I.C., van Hattum, T., van Riel, J.W., De Baere, K., Kempen, I. *et al.* (2020) Effects of on-farm and traditional hatching on welfare, health, and performance of broiler chickens. *Poultry Science* 99, 4662–4671. doi: 10.1016/j.psj.2020.06.052.

De Maesschalck, C., Eeckhaut, V., Maertens, L., De Lange, L., Marchal, L. *et al.* (2015) Effects of xylo-oligosaccharides on broiler chicken performance and microbiota. *Applied and Environmental Microbiology* 81, 5880–5888.

Ghazi, S., Rooke, J.A., Galbraith, H. and Bedford, M.R. (2002) The potential for the improvement of the nutritive value of soya-beanmeal by different proteases in broiler chicks and broiler cockerels. *British Poultry Science* 43, 70–77.

Hauschild, L., Bueno, C.F.D., Remus, A., de Paula Gobi, J., Isola, R.D.G. and Sakomura, N.K. (2015) Multiphase feeding program for broilers can replace traditional system. *Scientia Agricola* 72, 210–214.

Hesselman, K. and Aman, P. (1986) The effect of ß-glucanase on the utilization of starch and nitrogen by broiler chicks fed on barley of low or high viscosity. *Animal Feed Science and Technology* 15, 83–93.

Huang, K.H., Kemp, C. and Fisher, C. (2011) Effects of nutrition on water intake and litter moisture in broiler chickens. In: *Proceedings of the Australian Poultry Science Symposium*, Sydney, Australia, pp 26–31.

Huo, G.C., Fowler, V.R., Inborr, J. and Bedford, M.R. (1993) The use of enzymes to denature antinutritive factors in soybean. *Proceedings of the 2nd International Workshop on ANFs in Legume Seed*, p.60, Wageningen, The Netherlands.

Iji, P.A., Saki, A.A. and Tivey, D.R. (2001) Intestinal structure and function of broiler chickens on diets supplemented with a mannan oligosaccharide. *Journal of the Science of Food and Agriculture* 81, 1186–1192.

Jang, I.S., Ko, Y.H., Kang, S.Y. and Lee C.Y. (2007) Effect of commercial essential oils on growth performance, digestive enzyme activity and intestinal microflora population in broiler chickens. *Animal Feed Science and Technology* 134, 304–315.

Jensen, L.S., Fry, R.E., Alred, J.B. and Mcginnis, J. (1957) Improvement in the nutritional value of barley for chicks by enzyme supplementation. *Poultry Science* 36, 919–921.

Khattak, F.M., Pasha, T. N., Hayat, Z. and Mahmud, A. (2006) Enzymes in poultry nutrition *Journal of Animal and Plant Science* 16, 1–7.

Lemme, A. (2003) The "Ideal Protein Concept" 4 in broiler nutrition 2. Experimental data on varying dietary Ideal Protein levels. *AminoNews™* 4, 7–14.

Morgan, A.J. and Bedford, M.R. (1995) Wheat-specific feed enzymes. *Feed Compounder* 15, 25–27.

Moss, A.F., Chrystal, P.V. and Cadogan, D.J. (2020) Precision feeding enhances feed efficiency and carcass yield compared to broilers offered standard feeding programs. In: *Proceedings of the 2020 International Poultry Science Forum*, North American Meat Institute, Atlanta, GA, p. 32.

Moss, A.F., Chrystal, P.V., Cadogan, D.J., Wilkinson, S.J., Crowley, T.M. and Choct, M. (2021) Precision feeding and precision nutrition: a paradigm shift in broiler feed formulation? *Animal Bioscience* 34, 354–362. doi: 10.5713/ab.21.0034.

Noy, Y. and Sklan, D. (1996) Uptake capacity in vitro for glucose, methionine and in situ for oleic acid in the proximal small intestine of post-hatch chicks. *Poultry Science* 75, 998–1002.

Noy, Y. and Sklan, D. (1998) Are metabolic responses affected by early nutrition? *Journal of Applied Research* 7, 437–451.

Onrust, L., Baeyen, S., Haesebrouck, F., Ducatelle, R. and Van Immerseel, F. (2020) Effect of in feed administration of different butyrate formulations on *Salmonella* Enteritidis colonisation and cecal microbiota in broilers. *Veterinary Research* 51, 56–70.

Pinchasov, J. and Noy, Y. (1993) Comparison of posthatch holding time and subsequent early performance of broiler chicks and turkey poults. *British Poultry Science* 34,111–120.

Pirgozliev, V., Oduguwa, O., Acamovic, T. and Bedford, M.R. (2007) Diets containing *Escherichia coli*-derived phytase on young chickens and turkeys: Effects on performance, metabolisable energy and endogenous secretions and intestinal morphology. *Poultry Science* 86, 705–713.

Ribeiro, T., Cardoso, W., Ferreira, L.M.A., Lordelo, M.M.S, Coelho, E. *et al.* (2018) Xylo-oligosaccharides display a prebiotic activity when used to supplement wheat or corn-based diets for broilers. *Poultry Science* 97, 4330–4341.

Rota, C., Carraminana, J.J., Burillo, J. and Herrera, A. (2004) In vitro antimicrobial activity of essential oils from aromatic plants against selected foodborne pathogens. *Journal of Food Protection* 67, 1252–1256.

Savage, G.P. (1989) Antinutritional factors in peas. In: (J. Huisman, T. F.B. van der Poel and I.E. Liener, eds) *Recent Advances of Research in Antinutritional Factors in Legume Seeds: Proceedings of the First International Workshop on Antinutritional Factors (ANF) in Legume seeds.* Available at: https://edepot.wur.nl/313366 (accessed 29 March 2023).

Shang, Y., Kumar, S., Thippareddi, H. and Kyun Kim, W. (2018) Effect of dietary fructooligosaccharide (FOS) supplementation on ileal microbiota in broiler chickens. *Poultry Science* 97, 3622–3634.

Sharma, N.K., Creswell, D. and Swick, R.A (2014) Effect of feeding whole wheat and cracked corn on performance and carcass yield of broilers. In: *Proceedings of 14th European Poultry Conference*, Stavanger, Norway.

Singh, A.K., Mishra, B., Bedford, M.R. and Jha, R. (2021) Effects of supplemental xylanase and xylooligosaccharides on production performance and gut health variables of broiler chickens. *Journal of Animal Science and Biotechnology* 12, 98.

Slominski, B.A., Meng, X., Jia, W., Guenter, W. and Jones, O. (2004) The effect of lipase, amylase and protease addition on growth performance and nutrient digestion in young broiler chickens. Available at: https://www.cabi.org/Uploads/animal-science/worlds-poultry-science-association/WPSA-italy-2006/10641.pdf (accessed 29 March 2023).

Tan, J., McKenzie, C., Potamitis, M., Thorburn, A.N., Mackay, C.R. and Macia, L. (2014) The role of short-chain fatty acids in health and disease. *Advances in Immunology* 121, 91–119.

Van Harn, J., Dijkslag, M. and Van Krimpen, M. (2019) Effect of low protein diets supplemented with free amino acids on growth performance, slaughter yield, litter quality, and footpad lesions of male broilers. *Poultry Science* 98, 4868–4877.

Velasco, S., Ortiz, L.T., Alzueta, C., Rebolé, A., Treviño, J. and Rodríguez, M.L. (2010) Effect of inulin supplementation and dietary fat source on performance, blood serum metabolites, liver lipids, abdominal fat deposition, and tissue fatty acid composition in broiler chickens. *Poultry Science* 89, 1651–1662.

Windisch, W., Schedle, K., Plitzner, C. and Kroismayr, A. (2008) Use of phytogenic products as feed additives for swine and poultry. *Journal of Animal Science* 86, E140–E148.

Witjes, V.L., Bruckmaier, R.M., Gebhardt-Henrich, S.G. and Toscano, M.J. (2022) Effects of on-farm hatching on short term stress indicators, weight gain, and cognitive ability in layer chicks. *Applied Animal Behaviour Science* 254, 105692.

Xiao, C., Bao, G. and Hu, S. (2009) Enhancement of immune responses to Newcastle disease vaccine by a supplement of extract of *Momordica cochinchinensis* (Lour) Spreng seeds. *Poultry Science* 88, 2293–2297.

Xu, Z.R., Hu, C.H., Xia, M.S., Zhan, X.A. and Wang M.Q. (2003) Effects of dietary fructooligosaccharide on digestive enzyme activities, intestinal microflora and morphology of male broilers. *Poultry Science* 82, 1030–1036.

Yadav, J.L. and Sah, R.A. (2006) Supplementation of corn-soybean based layers diets with different levels of acid protease. *Journal of the Institute of Agriculture and Animal Science* 27, 93–102.

Yang, X., Yin, F., Yang, Y., Lepp, D., Yu., H. *et al.* (2018) Dietary butyrate glycerides modulate intestinal microbiota composition and serum metabolites in broilers. *Scientific Reports* 8, 4940.

Zhenping, S., Wenting, L., Ruikui, Y., Jia, L., Honghong, L. *et al.* (2013) Effect of a straw-derived xylooligosaccharide on broiler growth performance, endocrine metabolism, and immune response. *Canadian Journal of Veterinary Research* 77, 105–109.

Zhou, J., Wu S., Qi, G., Fu, Y., Wang, W. *et al.*(2021) Dietary supplemental xylooligosaccharide modulates nutrient digestibility, intestinal morphology, and gut microbiota in laying hens. *Animal Nutrition* 7, 152–162.

Zou, X., Ji, J., Qu, H., Wang, J., Shu, D.M. *et al.* (2019) Effects of sodium butyrate on intestinal health and gut microbiota composition during intestinal inflammation progression in broilers. *Poultry Science* 98, 4449–4456.

Zuidhof, M.J., Fedorak, M.V., Ouellette, C.A. and Wenger, I.I. (2017) Precision feeding: innovative management of broiler breeder feed intake and flock uniformity. *Poultry Science* 96, 2254–2263.

CHAPTER 5
Role of Pre, Pro and Synbiotics in Reducing Zoonotic Pathogen Abundance, Lowering Antimicrobial Resistance and Improving Food Safety: Old Concepts with a New Perspective

Farina Khattak* and Salvatore Galgano

Monogastric Science Research Centre, SRUC, Edinburgh, UK

Abstract

Zoonotic diseases are a multifaceted beast whether bacterial, viral or fungal; they spread to people through contact with animals carrying the pathogen. An estimated 60% of human infections and at least 75% of emerging diseases are zoonotic. The World Health Organization and the World Organisation for Animal Health agree that controlling zoonoses at their animal source is the most effective and economical way of protecting public health. Since the removal of antibiotics at the sub-therapeutic level, several strategies envisaging the use of probiotics and prebiotics, or a combination of them (i.e. synbiotics), as feed additives in poultry nutrition are gaining popularity. They are found to be effective in maintaining resilient performance without medication, consequently, a key subject that led to the optimization of different types of strategies, showing increased zootechnical performance, reduced pathogen growth and maintaining animal welfare. Nevertheless, while there is emerging evidence that probiotics can help reduce foodborne pathogens, studies link this approach to antimicrobial resistance rate owing to selection pressure, gene reservoir mechanisms and horizontal gene transfer. The main genus of interest in this chapter is *Campylobacter*. The aim is to review the existing literature, focusing on the application of non-antibiotic nutritional strategies in poultry,

*Corresponding author: Farina.khattak@sruc.ac.uk

DOI: 10.1079/9781800622746.0005

including probiotics or prebiotics administered alone or in combination, cell-free supernatant, bacteriophage therapy and other non-organic molecules such as peracetic acid, to combat pathogen colonization pre-slaughter. While using established disease challenge models, the mode of action of existing probiotics can be broadly described in three ways, including, but not limited to, competitive exclusion, antagonism and stimulation of the host immune response. Although prebiotics are relatively easy to add to feeds, there are many obstacles to their utilization, mainly due to in-feed stability and viability when delivered through the gastrointestinal tract to trigger a visible response both at the microbiota and host level. European studies indicate that on-farm non-antibiotic interventions could effectively lead to a 1.0 $\log_{10}$ *Campylobacter* reduction in caeca/faecal/carcass of chickens, resulting in a 90% reduction in human infections, but the biological significance of this reduction in *Campylobacter* concentration and the etiopathogenesis of related zoonoses remain a question. We will provide historic perspectives and recent updates on current non-antibiotic approaches and their impact on *Campylobacter* presence associated with poultry, with the main interest in their impact on poultry production and human health.

5.1 Introduction

Zoonosis is an infectious disease of humans caused by pathogens (e.g. bacteria, viruses) of animal origin, either through direct contact or contaminated food, water or the environment. The World Health Organization (WHO) estimates zoonosis incurrence in almost one in ten people (600 million), after eating contaminated food, resulting in a global annual burden of 33 million disability-adjusted life years (DALYs) and 420,000 premature deaths (WHO, 2015). Additionally, food supply globalization leads to an increased risk factor worldwide, including antimicrobial resistance (AMR) linked with foodborne pathogens because of the inappropriate use of antimicrobials in humans, animals and plants. It is estimated that by 2050, 10 million lives will be at risk, and a cumulative US$100 trillion will be lost owing to the impact of AMR if no proactive solutions are taken (O'Neill, 2016; WHO, 2022). The most reported zoonoses in humans are campylobacteriosis and salmonellosis (EU, 2021). *Campylobacter* is a fastidious commensal organism routinely found in cattle, sheep, swine and poultry, which can survive in various environments, yet broiler meat is considered the most common host (Hermans *et al.*, 2012; Sasaki *et al.*, 2013). The epidemiological evidence indicates that *Campylobacter*-positive undercooked chicken and poultry products such as contaminated chicken liver are trivial sources of human infection (Mor-Mur and Yuste, 2010). However, the only recorded campylobacteriosis cases currently reported are nosocomial, indicating that the reported figures are the tip of the iceberg.

5.2 Avian Gastrointestinal Microbiota and Microbiome

The terms 'microbiota' and 'microbiome' are often interchangeable but they are not perfect synonyms. 'Microbiota' refers to bacteria, fungi, archaea, protozoa and viruses living in the gut, whereas the term 'microbiome' refers to the collective genomes of the microorganisms in a given environment (Marchesi and Ravel, 2015). Nutritional studies in poultry have shown clear connections between microbiota and host growth, showing that diet is the main driver in shaping microbial communities over time (Borda-Molina *et al*., 2018). Gut communities have both benefits and costs to the host, with the primary benefits being the competitive exclusion of pathogens, immune stimulation and programming and contributions to host nutrition (Dibner and Richards, 2005). In the distal gut (i.e. caeca and colon), the microbiota also produces energy and nutrients such as vitamins, amino acids and short-chain fatty acids (SCFAs) from the undigested feed, which eventually become available for the host (Gaskins *et al*., 2002). SCFAs also play a significant role in regulating gut health; indeed, enterocytes use mostly butyrate for energy production, which also has anti-inflammatory properties (Hamer *et al*., 2008). SCFAs also impede the invasion and colonization of pathogens by lowering gut pH and inhibiting the conversion of bile to secondary bile products (Christl *et al*., 1997; Ricke, 2003).

The chicken's gastrointestinal tract includes the crop, proventriculus, gizzard, duodenum, jejunum, ileum, caeca, colon and cloaca (Yeoman *et al*., 2012). Each section has different metabolic functions that shape the microbial community and vice versa. The chicken crop harbours 10^8 to 10^9 cfu/g bacteria, which is usually dominated by lactobacilli (Gong *et al*., 2007). Most bacteria in the gizzard (10^7–10^8/g) are *Lactobacillus*, *Enterococcus*, lactose-negative enterobacteria and coliform bacteria (Rehman *et al*., 2007). Among the small intestinal segments, the bacterial density is the lowest in the duodenum due to short passage time and a dilution of digesta by secreted bile (Shapiro and Sarles, 1949). The duodenal bacterial community mainly consists of clostridia, streptococci, enterobacteria and lactobacilli (Waite and Taylor, 2015). *Lactobacillus* dominates the ileum microbiota as the primary group (70%), followed by members of the family *Clostridiaceae* (11%), *Streptococcus* (6.5%) and *Enterococcus* (6.5%) (Lu *et al*., 2003). Compared to the ileum, the caecum has a more diversified and stable microbial community, reported to be around 10^{10}–10^{11}/g of bacteria, including anaerobes (Videnska *et al*., 2013).

The chicken gut microbiota changes from crop to colon, and these changes are affected by diet (Torok *et al*., 2008), gender (Lumpkins *et al*., 2008), genotype (Zhao *et al*., 2013), housing condition (Nordentoft *et al*., 2011), floor litter (Torok *et al*., 2009), feed restriction (Callaway *et al*., 2009), stocking density (Guardia *et al*., 2011) and age (Corrigan *et al*., 2015). For example, *Firmicutes* are dominant in young chickens, while the *Bacteroidetes* are most common in adult birds (Callaway *et al*., 2009). In layers, four different profiles of caecal microbiota have been identified from the day of hatching until 60 weeks of age (Videnska *et al*., 2014).

Campylobacter is not simply a food contaminant but a member of the chicken gut microbiota and, as such, strategies targeting its reduction ought to consider ubiquitous microbiota modulation (Hermans *et al*., 2011). Currently, many strategies have been tested, targeting microbiota modulation, such as probiotics, prebiotics and faecal transplantation (Roberfroid *et al*., 2010; Fuentes and de Vos, 2016; de Cesare *et al*., 2019) as alternative antimicrobial strategies, which produce beneficial effects for the host whilst controlling zoonotic agents such as *Campylobacter*.

AMR amongst pathogens has been classified as a significant public health threat (Hofer, 2018). One of the main culprits for the increasing and persistent AMR spread is the misuse and overuse of antimicrobials (Dadgostar, 2019), for example, the administration of antimicrobials as growth promoters (AGPs), which is linked to increased performance and selection of resistant bacterial species (Cardinal *et al.*, 2020). Before AGPs were banned in most countries, they were commonly administered to birds; however, their mode of action towards increased performance had not been fully understood. It has been reported that some antimicrobials tend to enrich butyrate and lactate producers, probably explaining in part the effect on production (Robinson *et al.*, 2019). AMR distribution in livestock, including the poultry industry, not only represents a dangerous reservoir for the contamination and spread of AMR genes through the environment (Osman *et al.*, 2018) but also represents a risk factor associated with the passage of resistant bacteria to humans (Bortolaia *et al.*, 2016). This is especially important in the context of zoonosis and horizontal gene transfer (Nhung *et al.*, 2017), likely to be allowing the incorporation of AMR genes from zoonotic pathogens into a member of the human gut microbiota (Sørensen *et al.*, 2001), and when considering *Campylobacter*'s possible resistance to fluoroquinolones, erythromycin, clindamycin, kanamycin and ampicillin (Luangtongkum *et al.*, 2006).

Several strategies have been adopted to reduce the spread of AMR in poultry production, including increased antimicrobial surveillance, decreased use of antimicrobials, better management strategies and finding alternatives to antimicrobials (Apata, 2009). The following sections describe the current most common antimicrobial alternatives used to lower *Campylobacter* concentration *in vivo* in poultry.

5.3 Antimicrobial Alternatives Against *Campylobacter*

5.3.1 Probiotics

Probiotics are defined as 'live microorganisms, which, when administered in adequate amounts, confer a health benefit on the host' (FAO and WHO, 2001). This therapeutical approach is characterized by the administration of living microorganisms with a known beneficial effect on the host. Probiotics have been isolated from various sites, such as the gut and traditional fermented foods, and mostly belong to genera *Lactobacillus* and *Bifidobacterium*, although there are also some members of *Bacillus* and the yeast *Saccharomyces*, among others (Chaudhari and Dwivedi, 2022). Probiotics are thought to have several advantages, such as gut colonization whilst the probiotic is metabolically active, host health improvement either directly or indirectly by modulating other symbionts, relatively cheap production costs and safety given by the control over the strains used (Gaggìa *et al.*, 2010). Probiotics can be administered at different stages of the chicken's life following both temporal and longitudinal microbiota dynamics (Stanley *et al.*, 2014). They can directly affect health and performance, for example, by influencing cellulose and fat digestibility (Rubio *et al.*, 2015). The main mechanisms of action of probiotics can be summarized in the modulation of the microbial communities and production of active substances, such as bacteriocins and organic molecules, the influence of the immune function and interaction with the intestinal mucosa, aggregation with pathogens and competition for nutrients (Tiwari *et al.*, 2012). Owing to these reasons, probiotics could modulate pathogens' establishment and spread and are therefore considered antimicrobial alternatives (Barba-Vidal *et al.*, 2019). Many studies have therefore focused their attention on the effect of different probiotic strains on *Campylobacter* (Saint-Cyr *et al.*, 2016; Deng *et al.*, 2020). However, probiotic beneficial effects are not only limited to lowering *Campylobacter* levels in the avian gut but are also extended to reducing environmental contamination at farms and ultimately contributing to enhanced carcass hygienic indices (Smialek *et al.*, 2018).

In a study from 2010, *Lactobacillus plantarum* PCS20 and *Bifidobacterium longum* PCB133 were independently tested in broilers, resulting in evidence showing that *B. longum*, but not *L. plantarum* was successfully found in excreta after ~21 days and was shown to significantly reduce *C. jejuni* (Santini *et al.*, 2010). Another study indicated, however, that several *Lactobacillus* species, such as *L. salivarius, L. johnsonii, L. reuteri, L. crispatus and L. gasseri*, were not only able to inhibit the growth of *C. jejuni* but also reduced the expression of *C. jejuni* virulence-related genes (apart from *L. reuteri*) and increased macrophage activity (Taha-Abdelaziz *et al.*, 2019). These findings probably confirm what is known as probiotic strain selectivity, in which probiotics are species/strain specific toward definite pathogen strains within the appropriate eukaryotic cell-type environment (Wine *et al.*, 2009). Interestingly, whole-genome-sequencing analysis of *L. salivarius* SMXD51 revealed the presence of several genes known to be related to gastrointestinal survival and adherence, for example, genes coding for acid tolerance and aggregation-promoting factor or adhesion capacity (Kergourlay *et al.*, 2012), corroborating the importance of competition mechanisms to combat *Campylobacter* adhesion.

5.3.2 Prebiotics

Prebiotics are defined as selectively fermented ingredients that result in specific changes in the composition and/or activity of the gastrointestinal microbiota, thus conferring benefits upon host health (Gibson *et al.*, 2010). Some commonly used prebiotics are fructo-oligosaccharides (FOS), mannan-oligosaccharides (MOS), galacto-oligosaccharides (GOS), inulin, lactulose and xylo-oligosaccharides (XOS). In addition, prebiotics modulate the microbiota towards the proliferation of species possibly beneficial for the host, for example, digesting dietary molecules otherwise usually escaping host digestion (Zimmermann *et al.*, 2001). Several plant-derived substrates have been proposed and tested as a prebiotic, such as resistant starch or small undigested oligosaccharides, which seemed to affect the caecal concentration of bacterial SCFAs (Pan *et al.*, 2009). As well as the induction of SCFA production, other mechanisms of action of prebiotics are the contribution to pH homeostasis, lipid metabolism control, immunomodulation via binding the G protein-coupled receptor of the gut-associated lymphoid tissue or via interaction between some SCFAs and cytokines production, controlling inflammation and improving nutritional absorption (Gowrishankar *et al.*, 2021). For these reasons, prebiotics have been studied widely in the context of limiting *Campylobacter* concentration within the chicken gastrointestinal tract, as already reviewed (Kim *et al.*, 2019). Several studies have reported that using prebiotics alone did not reduce *Campylobacter* concentration. Indeed, an MOS-based prebiotic was linked to increased villi surface area and reduction of *Salmonella* levels but not *Campylobacter* in turkeys (Rahimi *et al.*, 2019). Similarly, GOS given to challenged birds did not reduce *Campylobacter*, but was linked to increased performance (Flaujac Lafontaine *et al.*, 2020).

In contrast, a *Saccharomyces*-derived prebiotic was associated with a decreased level of *Campylobacter* in the caeca, increased body weight, average daily gain and average daily feed intake (Froebel *et al.*, 2019), and 2 g/kg MOS reduced ileal and caecal *C. jejuni* concentration at days 24 and 42 in broilers, whilst also increasing *Lactobacillus* in two locations and time points (Rostami *et al.*, 2022). These findings might suggest that prebiotics are unlikely to exert a direct effect on *Campylobacter*; however, their likely indirect effect could be mediated by specific microbial communities and their related molecules. Following this rationale, prebiotics that can lead to an enrichment of those populations with the known detrimental outcome on *Campylobacter* could be successfully applied to reduce *Campylobacter* levels *in vivo* in poultry production to reduce the zoonotic risk pre-harvest.

5.3.3 Synbiotics

The synergic combinations of probiotics and prebiotics, in which the latter selectively provides advantages to the former, are defined as synbiotics (Pandey *et al.*, 2015). As already reviewed by Malik *et al.* (2019), there are many examples of the beneficial effects of synbiotics both on pathogen reduction and improved performance. For example, the combination of *B. longum* and GOS had been reported to decrease *C. jejuni* abundance in excreta samples (Baffoni *et al.*, 2012), whereas a synbiotic based on *B. longum* and XOS reportedly resulted in a reduction of *C. jejuni* in caeca of broilers after continuous administration (Baffoni *et al.*, 2017). Importantly, synbiotics have been successfully tested against other zoonotic agents; for example, FOS used with *L. reuteri*, *Enterococcus faecium*, *Bifidobacterium animalis* and *Pediococcus acidilactici* was able to exert a beneficial effect against *S. enterica* infection in caeca of layers, whilst also increasing body weight (Luoma *et al.*, 2017). This evidence shows that probiotic strains and molecules that improve functional microbiota modulation could perform better than administering a single component based on the synergy of action.

5.3.4 Bacteriophages

Bacteriophages are intracellular viruses specifically infecting bacteria, with two main typologies: virulent phages characterized by a lytic life cycle (i.e. the formation and liberation of virions via cell-lysis) and temperate phages with a lysogenic life cycle, through which viral DNA is incorporated in the bacterial chromosome (Clavijo and Flórez, 2017). Bacteriophages can therefore be used as a therapeutical approach to target *Campylobacter*, leading to its reduction *in vivo* (Zhang *et al.*, 2022). Different strategies have been tested so far, including single phage therapy (Jeon *et al.*, 2012), phage cocktails (Molina *et al.*, 2021), a phage–antibiotic combination (Tagliaferri *et al.*, 2019) or phage-derived enzymes such as depolymerases (Pires *et al.*, 2016) and endolysins (Schmelcher and Loessner, 2021). Both single- and cocktail-phage approaches successfully induced up to $\log_{10}$ CFU/g *C. jejuni* concentration reduction between one and four weeks after treatment in broiler birds (Fischer *et al.*, 2013). Other authors corroborated these findings by describing a reduction in *C. jejuni* abundance in broilers after oral treatment with *Campylobacter*-specific phage cocktail, importantly without further affecting microbiota species (Richards *et al.*, 2019), or showing a 2 log reduction in *C. jejuni* and *C. coli* counts in excreta of birds artificially exposed to *Campylobacter* spp. (Carvalho *et al.*, 2010).

5.3.5 Avian antimicrobial peptides

Avian antimicrobial peptides (AAPs) are small molecules of ~50 amino acids, part of the innate immunity of poultry, with cathelicidins and β-defensins representing the main families; their main mechanism of action is exerted either by their hydrophobicity, leading to cell lysis via formation of transmembrane pores and ion channels, or by the recruitment and activation host-immune cells (Nguyen *et al.*, 2021). Chicken β-defensins (e.g. AvBD1–14) show tissue-specific expression levels, with high mRNA levels of AvBD4 to AvBD7 found in chicken bone marrow but not in leukocyte extracts, and they have known activity towards Gram-positive and Gram-negative bacteria, and fungi, such as *C. jejuni*, *C. perfringens*, *S. aureus*, *C. albicans* and *S. cerevisiae* (van Dijk *et al.*, 2008). Cathelicidins have a critical role in the innate avian immune system; they are mainly expressed by epithelial cells or bone marrow (Wang *et al.*, 2020), and are grouped into four classes, CATH-1, CATH-2, CATH-3 and CATH-B1, with the truncated peptides CATH-1(6–26), CATH-2(1–15) showing less toxicity (Nguyen *et al.*, 2021).

The immunological role of AAPs in response to *Campylobacter* has been demonstrated by several authors, indicating their potential therapeutic efficacy, for example, as a feed additive upon *in vitro* production of a recombinant. During a challenge trial with different concentrations of *C. jejuni*, it was demonstrated that there was a differential expression level of some β-defensins, apparently dependent on the initial inoculum concentration; for example, duodenal AvBD10 was up-regulated in birds given 10^6 CFU but down-regulated in the ileum of the 10^8 CFU group (Garcia *et al.*, 2018). Interestingly, it was demonstrated that *Campylobacter* challenge was significantly associated with reduced expression levels of some β-defensins (i.e. AvBD3, AvBD4, AvBD8, AvBD13 and AvBD14) and cathelicidins (i.e. CTHL2 and CTHL3), when compared to *Salmonella*-challenged birds, probably indicating a different host-immune response to both bacteria with a possible explanation for the high colonization level of *Campylobacter* (Meade *et al.*, 2009). The importance of AAPs in intestinal resistance to *Campylobacter* colonization was underlined in a study assessing the expression levels of genes in chicken genetic lines resistant or susceptible to *Campylobacter* colonization, emphasized by the caecal up-regulation of both β-defensins 10 and 12 in resistant birds (Li *et al.*, 2010).

5.3.6 Bacteriocins

Bacteria also produce antimicrobial peptides, called bacteriocins, produced by Gram-positive and Gram-negative bacteria to inhibit the growth of other microorganisms (Benítez-Chao *et al.*, 2021). In nature, bacteriocins have several roles, such as assisting colonization by inhibiting competing microbes in an already occupied microenvironment, functioning as signalling peptides targeting both bacteria (e.g. quorum sensing) and the host's immune cells (Mohan, 2015). For these reasons, like AAP, bacteriocins represent rather interesting antimicrobial alternative molecules, which could be synthesised *in vitro* and the administration of which could either hinder *Campylobacter* growth pre-harvest or modulate the rest of the microbial communities, leading to beneficial effects for the host.

As already reviewed by Svetoch and Stern (2010), therapeutic bacteriocins could reduce poultry *Campylobacter* levels in the caeca of chickens from $>10^8$ CFU/g to non-detectable or very low levels. In another study, bacteriocins produced by *Bacillus circulans* were able to reduce *Campylobacter* concentration to non-detectable from ~1 x 10^6 in the caeca of challenged chickens (Cole *et al.*, 2006). Corroborating these results, two different experiments demonstrated that bacteriocins derived from *L. salivarius* added in-feed reduced caecal *Campylobacter* concentration (Stern *et al.*, 2006), also showing that 250 mg of purified bacteriocins/kg of feed was able to reduce pathogen concentration otherwise not affected by the administration of the probiotic-producing bacteriocins only (Stern *et al.*, 2008).

Similar molecules to bacteriocins and AAPs are produced by other groups of organisms, such as plants (thionins), e.g. compounds in the leaf of *Ferula gummosa* are active towards *C. jejuni* (Kafi *et al.*, 2022), amphibians (magainins) or insects (melittin) (Karpinski and Szkaradkiewicz, 2013). However, this chapter does not describe their applications within the topic of reducing zoonosis.

5.3.7 Cell-free supernatant

There has been considerable evidence showing that the protective role of probiotics was not only dependent on viable bacteria (Harb *et al.*, 2013). Indeed, in addition to the aforementioned bacteriocins, there is a plethora of active extracellular antimicrobial molecules of bacterial origin, such as hydrogen peroxide, reuterin, nisin and pediocin, that can act synergistically to lead to effective microbial inhibition. This suggests the use of cell-free supernatant (CFS, i.e. the ensemble of all the bacterial molecules found in the supernatant after centrifugation) as an antimicrobial alternative (Beristain-Bauza *et al.*, 2016). The technical steps to produce CFS are relatively easy and include the propagation of the probiotic strain(s) of interest, the separation of the cellular component by centrifugation followed by filter sterilization of the supernatant and, finally, the lyophilization of the CFS for downstream application (Kim *et al.*, 2021).

In one study, CFS from *Lactobacillus fermentum* created a clear inhibition zone in a disc diffusion test on *C. coli* and *C. jejuni*, whereas *L. thermotolerans* CFS was effective only on *C. coli* and *L. reuteri* CFS was only effective on one strain of *C. jejuni* (Bratz *et al.*, 2015). The same study also showed no activity towards *Campylobacter* spp. of CFS from *E. faecium* or 13 *Bifidobacterium* strains tested. These findings clearly indicate specificity between the CFS molecules from specific probiotics and pathogens, probably pointing towards target-recognition-like mechanisms. The ability to use this therapeutic approach by targeting specific species is interesting and is accompanied by great potential for applications in both the poultry industry and human medicine. Conversely, a similar study assessed the growth of *C. coli* via optical density measurement whilst growing in incubation with different CFS concentrations from similar probiotic strains and demonstrated an inhibition at a concentration from 1:1 to 10:1 of *E. faecium*, *B. animalis*, *P. acidilactici* and *L. reuteri* (Mortada *et al.*, 2020). The findings from these two experiments seem to be in contrast and indicate that further research is needed, especially to classify the CFS molecular composition in correlation to its effect on *Campylobacter*. Moreover, CFS as a feed additive would represent a novel product reducing zoonotic load pre-harvest, thus improving gut health and lowering antibiotic usage and AMR spread.

5.3.8 Non-organic alternatives

Within this category, we describe molecules used *in vivo,* as a biocide or as a broad-spectrum antimicrobial alternative, which could help reduce the *Campylobacter* concentration pre-harvest, reducing the potential campylobacteriosis risk. For example, in-feed ferric tyrosine (TYPLEX® Chelate) successfully inhibited *C. jejuni* biofilm formation and reduced both *C. jejuni* and *E. coli* colonization in broiler chickens at market age. (Khattak *et al.*, 2018). In another study, peracetic acid (PAA), commonly used for its biocide effects (Kitis, 2004), was tested in broilers. PAA was generated via the administration of two precursors, sodium percarbonate (SP) and tetraacetylethylenediamine (TAED, Aga2 Tech), within physiological pH values, and showed antimicrobial properties *in vivo*, without releasing harmful by-products (Galgano *et al.*, 2023). Furthermore, triclosan, benzalkonium chloride, cetylpyridinium chloride, chlorhexidine diacetate and trisodium phosphate are amongst the non-organic molecules investigated for their action on *Campylobacter*. It has been shown, however, that *Campylobacter* could develop efflux pump-mediated resistance to these compounds (Mavri and Smole Možina, 2013). The mode of action of non-organic alternatives is intervention dependent; for example, the precursor-derived PAA was formed by hydrolysis of SP and TAED in the upper gut and, once active, interacted with strong oxidizing activity towards bacterial membranes, leading to bacterial inactivation (Galgano *et al.*, 2021).

5.4 *In vitro* Investigation of the Gut Microbiota

Depending on the research question and the investigation set-up, the different approaches to studying the interactions and composition of gut microbial communities are generally classified as culture-based methods, culture-independent methods and mathematical modelling.

5.4.1 Culture-based methods

Culture-based techniques have many applications, including new species identification, dynamics exploration within experimental environments (Lagier *et al.*, 2015) and bacterial quantification (Khattak *et al.*, 2022). However, they require a strict set-up, e.g. temperature, pO_2 and medium composition, specific for the tested species, or similar to chyme composition if the study is aimed at assessing microbiota dynamics within a specific gut location (Bindelle *et al.*, 2007). Culture-based experiments can be either in batch (Sánchez *et al.*, 2006) or in continuous systems (Tanner *et al.*, 2014), with a growing level of complexity in the latter, which allow the *in vitro* simulation of parameters such as retention time and therefore allow temporal dynamics closer to physiological ones (Galgano *et al.*, 2018). In addition, complex bioreactor types have been designed and tested, allowing the simulation of other factors, such as the interaction between microbiota and host cells (Zhou *et al.*, 2018).

5.4.2 Culture-independent methods

The entire microbiota composition can be explored with metataxonomic methods (Marchesi and Ravel, 2015), which could be based on the sequencing of marker genes such as one of the 16S rRNA gene regions (Chakravorty *et al.*, 2007), or on the sequencing of the whole genome (Medvecky *et al.*, 2018), with increasing level of details in the output and increasing costs.

5.4.3 Mathematical modelling

The typical bacterial growth curve, from lag to decline phase, can be modelled *in silico* (i.e. computationally) using specific equations (Monod, 1949), which could be used in the context of a complex system of ordinary differential equations, generating simulations and predictions of the microbiota dynamics through experimental conditions (Kettle *et al.*, 2018; Galgano *et al.*, 2020). Mathematical models of gut microbiota are reliable computational tools that can anticipate changes caused by microbiota modulation, such as the administration of probiotics or prebiotics (Esser *et al.*, 2015) and therefore represent a valuable instrument in the study of the interaction of different intervention types such as the ones described in this chapter.

5.5 *In vivo* Testing Methods

Reliable *in vivo* infection models are required to test the efficacy of, for example, novel products for the inhibition of caecal colonization by *Campylobacter* in broilers. Three different infection models, 'natural', 'seeded litter', and 'gavaging' methods, have been successfully and reliably used to colonize 21-day-old broilers by 28 days of age with *Campylobacter* spp. (Khattak *et al.*, 2018, 2021). The natural infection model involves using *Campylobacter*-positive litter from commercial farms/external trials, facilitating the faecal/oral challenge route (Khattak *et al.*, 2017). However, the limitation of this method is the timebound need for outsourcing litter and the risk of unequal pathogen distribution within a pen. The seeded litter challenge model was therefore developed and used to horizontally transfer *C. jejuni* in birds within an experimental pen via inoculating the litter with a mixture of *Campylobacter* species and excreta in a specific medium, mimicking infection dynamics in commercial settings (Sandilands *et al.*, 2018). However, this procedure does not allow precise *Campylobacter* quantification per bird. In contrast, the oral gavaging method allows the administration of a precise amount of *Campylobacter* inoculum directly into the crop of chickens (Arsi *et al.*, 2015). Nevertheless, this technique requires a high level of expertise of trained personnel, accompanied by increased costs due to the individual manipulations of all the birds in the flock.

5.6 Pro and Prebiotics: Marketing Trends and Future Forecasts

The market for probiotics and prebiotics has seen a rapid and constant increase throughout the years since the launch of the first products (e.g. Yakult® and non-digestible fibres) through the 20th century.

The global probiotic market is segmented by product type, that is, food and beverages, dietary supplements and animal feed 'biotics'. According to the global market forecasts, the probiotic market is expected to grow at a compound annual growth rate (CAGR) of 7.5% from 2022 to 2030, with increased market share by food and beverages compared to animal feed (MAP, 2021a, b). An annual market growth (CGAR) of 14.9% from 2022 to 2030 is expected for prebiotics in the food and feed industry, mostly due to rising awareness regarding the benefits of functional probiotic foods, increased demand for high-nutrition and low-calorie processed food products, the ban of in-feed antibiotics in livestock, and the known probiotic-driven beneficial effects in animals.

Among all segments, the lowest global growth is recorded in the animal feed sector, probably because of the strict regulations, the need for further testing of novel feed additives and the lengthy approval processes.

5.7 Conclusion

The current evidence suggests that the use of probiotics, prebiotics and a combination thereof is a valid and much-needed approach to lower the sub-therapeutic need for antibiotics in animals and thus aids towards lowering AMR and, ultimately, enhances food safety. The concept of using 'biotics' to promote human health is old but determining how they affect foodborne pathogens is still in its infancy. Although the reliable and commercially relevant *in vivo* infection models for *Campylobacter* spp. challenge in broilers have been established, concerns regarding the optimal dosing and understanding the delicate interaction between *C. jejuni* and the chicken immune system remains a challenge for the scientists. Therefore, further targeted research is required to understand the full potential of these antimicrobial alternatives in the context of campylobacteriosis.

References

Apata, D.F. (2009) Antibiotic resistance in poultry. *International Journal of Poultry Science* 8, 404–408.

Arsi, K., Donoghue, A.M., Woo-Ming, A., Blore, P.J. and Donoghue, D.J. (2015) Intracloacal inoculation, an effective screening method for determining the efficacy of probiotic bacterial isolates against *Campylobacter* colonization in broiler chickens. *Journal of Food Protection* 78, 209–213.

Baffoni, L., Gaggìa, F., di Gioia, D., Santini, C., Mogna, L. and Biavati, B. (2012) A Bifidobacterium-based synbiotic product to reduce the transmission of *C. jejuni* along the poultry food chain. *International Journal of Food Microbiology* 157, 156–161.

Baffoni, L., Gaggìa, F., Garofolo, G., di Serafino, G., Buglione, E. *et al.* (2017) Evidence of *Campylobacter jejuni* reduction in broilers with early synbiotic administration. *International Journal of Food Microbiology* 251, 41–47.

Barba-Vidal, E., Martín-Orúe, S.M. and Castillejos, L. (2019) Practical aspects of the use of probiotics in pig production: A review. *Livestock Science* 223, 84–96.

Benítez-Chao, D.F., León-Buitimea, A., Lerma-Escalera, J.A. and Morones-Ramírez, J.R. (2021) Bacteriocins: an overview of antimicrobial, toxicity, and biosafety assessment by *in vivo* models. *Frontiers in Microbiology* 12, 630695.

Beristain-Bauza, S.C., Mani-López, E., Palou, E. and López-Malo, A. (2016) Antimicrobial activity and physical properties of protein films added with cell-free supernatant of *Lactobacillus rhamnosus*. *Food Control* 62, 44–51.

Bindelle, J., Buldgen, A., Boudry, C. and Leterme, P. (2007) Effect of inoculum and pepsin-pancreatin hydrolysis on fibre fermentation measured by the gas production technique in pigs. *Animal Feed Science and Technology* 132, 111–122.

Borda-Molina, D., Seifert, J. and Camarinha-Silva, A. (2018) Current perspectives of the chicken gastrointestinal tract and Its microbiome. *Computational and Structural Biotechnology Journal* 16, 131–139.

Bortolaia, V., Espinosa-Gongora, C. and Guardabassi, L. (2016) Human health risks associated with antimicrobial-resistant enterococci and *Staphylococcus* aureus on poultry meat. *Clinical Microbiology and Infection* 22, 130–140.

Bratz, K., Gölz, G., Janczyk, P., Nöckler, K. and Alter, T. (2015) Analysis of *in vitro* and *in vivo* effects of probiotics against *Campylobacter* spp. *Berliner und Munchener Tierarztliche Wochenschrift* 128, 155–162.

Callaway, T.R., Dowd, S.E., Wolcott, R.D., Sun, Y., McReynolds, J.L. *et al.* (2009) Evaluation of the bacterial diversity in cecal contents of laying hens fed various molting diets by using bacterial tag-encoded FLX amplicon pyrosequencing. *Poultry Science* 88, 298–302.

Cardinal, K.M., Pires, S. and Ribeiro, A.M.L. (2020) Growth promoter in broiler and pig production. *PUBVET* 14, 1–11.

Carvalho, C., Susano, M., Fernandes, E., Santos, S., Gannon, B. *et al.* (2010) Method for bacteriophage isolation against target *Campylobacter* strains. *Letters in Applied Microbiology* 50, 192–197.

Chakravorty, S., Helb, D., Burday, M., Connell, N. and Alland, D. (2007) A detailed analysis of 16S ribosomal RNA gene segments for the diagnosis of pathogenic bacteria. *Journal of Microbiological Methods* 69, 330–339.

Chaudhari, A. and Dwivedi, M.K. (2022) The concept of probiotics, prebiotics, postbiotics, synbiotics, nutribiotics, and pharmabiotics. In *Probiotics in the Prevention and Management of Human Diseases*, Elsevier, pp. 1–11.

Christl, S.U., Bartram, H.P., Paul, A., Kelber, E., Scheppach, W. and Kasper, H. (1997) Bile Acid Metabolism by Colonic Bacteria in Continuous Culture: Effects of Starch and pH. *Annals of Nutrition and Metabolism* 41, 45–51.

Clavijo, V. and Flórez, M.J.V. (2017) The gastrointestinal microbiome and its association with the control of pathogens in broiler chicken production: A review. *Poultry Science* 97, 1006–1021.

Cole, K., Farnell, M.B., Donoghue, A.M., Stern, N.J., Svetoch, E.A., Eruslanov, B.N. *et al.* (2006) Bacteriocins Reduce *Campylobacter* Colonization and Alter Gut Morphology in Turkey Poults. *Poultry Science* 85, 1570–1575.

Corrigan, A., de Leeuw, M., Penaud-Frézet, S., Dimova, D. and Murphya, R.A. (2015) Phylogenetic and functional alterations in bacterial community compositions in broiler ceca as a result of mannan oligosaccharide supplementation. *Applied and Environmental Microbiology* 81, 3460–3470.

Dadgostar, P. (2019) Antimicrobial resistance: implications and costs. *Infection and Drug Resistance* 12, 3903–3910.

de Cesare, A., Caselli, E., Lucchi, A., Sala, C., Parisi, A. *et al.* (2019) Impact of a probiotic-based cleaning product on the microbiological profile of broiler litters and chicken caeca microbiota. *Poultry Science* 98, 3602–3610.

Deng, W., Dittoe, D.K., Pavilidis, H.O., Chaney, W.E., Yang, Y. and Ricke, S.C. (2020) Current perspectives and potential of probiotics to limit foodborne *Campylobacter* in poultry. *Frontiers in Microbiology* 11, 583429.

Dibner, J.J. and Richards, J.D. (2005) Antibiotic growth promoters in agriculture: history and mode of action. *Poultry Science* 84, 634–643.

Esser, D.S., Leveau, J.H.J. and Meyer, K.M. (2015) Modeling microbial growth and dynamics. *Applied Microbiology and Biotechnology* 99, 8831–8846.

EU (2021) The European Union One Health 2019 Zoonoses Report. *EFSA Journal* 19, e06406.

FAO and WHO (2001) Probiotics in food health and nutritional properties and guidelines for evaluation. Available at: https://www.fao.org/3/a0512e/a0512e.pdf (accessed 6 January 2023).

Fischer, S., Kittler, S., Klein, G. and Glünder, G. (2013) Impact of a single phage and a phage cocktail application in broilers on reduction of *Campylobacter jejuni* and development of resistance. *PLoS One* 8, e78543.

Flaujac Lafontaine, G.M., Richards, P.J., Connerton, P.L., O'Kane, P.M., Ghaffar, N.M. *et al.* (2020) Prebiotic driven increases in IL-17A do not prevent *Campylobacter jejuni* colonization of chickens. *Frontiers in Microbiology* 10, 3030.

Froebel, L.K., Jalukar, S., Lavergne, T.A., Lee, J.T. and Duong, T. (2019) Administration of dietary prebiotics improves growth performance and reduces pathogen colonization in broiler chickens. *Poultry Science* 98, 6668–6676.

Fuentes, S. and de Vos, W.M. (2016) How to manipulate the microbiota: fecal microbiota transplantation. In: Schwiertz, A. (ed.) *Microbiota of the Human Body. Advances in Experimental Medicine and Biology*. Springer, Cham. https://doi.org/10.1007/978-3-319-31248-4_10

Gaggìa, F., Mattarelli, P. and Biavati, B. (2010) Probiotics and prebiotics in animal feeding for safe food production. *International Journal of Food Microbiology* 141, S15–S28.

Galgano, S., Kettle, H., Free, A., Olukosi, O.A. and Houdijk, J.G.M. (2018) Dynamic modelling of the core microbiota in the porcine gut. *Proceedings of the British Society of Animal Science* 039.

Galgano, S., Kettle, H., Free, A. and Houdijk, J.G.M. (2020) Dynamic modelling of pig gut microbiota. *Proceedings of the British Society of Animal Science* 11, 066.

Galgano, S., Mistry, D., di Maggio, F., Farthing, K., Dalby, N. *et al.* (2021) Peracetic acid precursors for non-antibiotic modulation of gut microbiota and possible modulation of performance in broilers. *British Poultry Abstracts* 17, 18–19.

Galgano, S., Conway, L., Maggio, F. di, Farthing, K., Dalby, N. *et al.* (2023) Precursor-derived in-water peracetic acid impacts on broiler performance, gut microbiota and antimicrobial resistance genes. *Poultry Science* 102, 102368.

Garcia, J.S., Byrd, J.A. and Wong, E.A. (2018) Expression of nutrient transporters and host defense peptides in *Campylobacter* challenged broilers. *Poultry Science* 97, 3671–3680.

Gaskins, H.R., Collier, C.T. and Anderson, D.B. (2002) Antibiotics as growth promotants: mode of action. *Animal Biotechnology,* 13, 29–42.

Gibson, G.R., Scott, K.P., Rastall, R.A., Tuohy, K.M., Hotchkiss, A. *et al.* (2010) Dietary prebiotics: current status and new definition. *Food Science & Technology Bulletin: Functional Foods* 7, 1–19.

Gong, J., Si, W., Forster, R.J., Huang, R., Yu, H. *et al.* (2007) 16S rRNA gene-based analysis of mucosa-associated bacterial community and phylogeny in the chicken gastrointestinal tracts: from crops to ceca. *FEMS Microbiology Ecology* 59, 147–157.

Gowrishankar, S., Kamaladevi, A. and Karutha Pandian, S. (2021) Prebiotics mechanism of action: an overview. In: *Advances in Probiotics*. Elsevier, pp. 137–148.

Guardia, S., Konsak, B., Combes, S., Levenez, F., Cauquil, L. *et al.* (2011) Effects of stocking density on the growth performance and digestive microbiota of broiler chickens. *Poultry Science* 90, 1878–1889.

Hamer, H.M., Jonkers, D., Venema, K., Vanhoutvin, S., Troost, F.J. and Brummer, R.J. (2008) Review article: the role of butyrate on colonic function. *Alimentary Pharmacology and Therapeutics* 27, 104–119.

Harb, H., van Tol, E.A.F., Heine, H., Braaksma, M., Gross, G. *et al.* (2013) Neonatal supplementation of processed supernatant from *Lactobacillus rhamnosus* GG improves allergic airway inflammation in mice later in life. *Clinical & Experimental Allergy* 43, 353–364.

Hermans, D., van Deun, K., Martel, A., van Immerseel, F., Messens, W. *et al.* (2011) Colonization factors of *Campylobacter jejuni* in the chicken gut. *Veterinary Research* 42, 1–14.

Hermans, D., Pasmans, F., Messens, W., Martel, A., van Immerseel, F. *et al.* (2012) Poultry as a host for the zoonotic pathogen *Campylobacter jejuni*. *Vector-Borne and Zoonotic Diseases* 12, 89–98.

Hofer, U. (2018) The cost of antimicrobial resistance. *Nature Reviews Microbiology* 17, 3–3.

Jeon, J., Kim, J., Yong, D., Lee, K. and Chong, Y. (2012) Complete genome sequence of the podoviral bacteriophage YMC/09/02/B1251 ABA BP, which causes the lysis of an OXA-23-producing carbapenem-resistant *Acinetobacter baumannii* isolate from a septic patient. *Journal of Virology* 86, 12437–12438.

Kafi, M.K., Bolvari, N.E., Pour, S.M., Moghadam, S.K., Karimi, E. and Oskoueian, E. (2022) Encapsulated phenolic compounds from Ferula gummosa leaf: A potential phytobiotic against *Campylobacter jejuni* infection. *Journal of Food Processing and Preservation* 46, e16802

Karpinski, T.M. and Szkaradkiewicz, A.K. (2013) Characteristic of bacteriocines and their applications. *Polish Journal of Microbiology* 62, 223–235.

Kergourlay, G., Messaoudi, S., Dousset, X. and Prévost, H. (2012) Genome sequence of lactobacillus salivarius SMXD51, a potential probiotic strain isolated from chicken cecum, showing Anti-*Campylobacter* activity. *Journal of Bacteriology* 194, 3008–3009.

Kettle, H., Holtrop, G., Louis, P. and Flint, H.J. (2018) microPop: Modelling microbial populations and communities in R. *Methods in Ecology and Evolution* 9, 399–409.

Khattak, F., Williams, S., Mahdavi, J. and Houdijk, J. (2017) Efficacy of TYPLEX(TM) chelate in broilers with natural *Campylobacter* challenge. *British Poultry Abstracts* 13, 1–39.

Khattak, F., Paschalis, V., Green, M., Houdijk, G.M., Soultanas, P. and Mahdavi, J. (2018) TYPLEX® Chelate, a novel feed additive, inhibits *Campylobacter jejuni* biofilm formation and cecal colonization in broiler chickens. *Poultry Science* 97, 1391–1399.

Khattak, F., Pedersen, N.R., Matthiesen, R. and Houdijk, J.G.M. (2021) Lacto-fermented rapeseed meal additive: a nutritional intervention to reduce *Campylobacter jejuni* colonisation and improve performance in broilers. *British Poultry Abstracts* 17, 7–8.

Khattak, F.I., Galgano, S.I. and Houdijk, J. (2022) Bacterial concentration and *Campylobacter* spp. quantification differ when fresh or ultra-frozen samples are analysed over time using molecular biology and culture-based methods. *PLoS One* 17, e0274682.

Kim, S.A., Jang, M.J., Kim, S.Y., Yang, Y., Pavlidis, H.O. and Ricke, S.C. (2019) Potential for prebiotics as feed additives to limit foodborne *Campylobacter* establishment in the poultry gastrointestinal tract. *Frontiers in Microbiology* 10, 91.

Kim, Y.J., Yu, H.H., Song, Y.J., Park, Y.J., Lee, N.K. and Paik, H.D. (2021) Anti-biofilm effect of the cell-free supernatant of probiotic *Saccharomyces cerevisiae* against *Listeria* monocytogenes. *Food Control* 121, 107667

Kitis, M. (2004) Disinfection of wastewater with peracetic acid: A review. *Environment International* 30, 47–55.

Lagier, J.C., Edouard, S., Pagnier, I., Mediannikov, O., Drancourt, M. and Raoult, D. (2015) Current and past strategies for bacterial culture in clinical microbiology. *Clinical Microbiology Reviews* 28, 208–236.

Li, X., Swaggerty, C.L., Kogut, M.H., Chiang, H.I., Wang, Y. *et al.* (2010) Gene expression profiling of the local cecal response of genetic chicken lines that differ in their susceptibility to *Campylobacter jejuni* colonization. *PLoS One* 5, e11827.

Lu, J., Idris, U., Harmon, B., Hofacre, C., Maurer, J.J. and Lee, M.D. (2003) Diversity and succession of the intestinal bacterial community of the maturing broiler chicken. *Applied Environmental Microbiology* 69, 6816–6824.

Luangtongkum, T., Morishita, T.Y., Ison, A.J., Huang, S., McDermott, P.F. and Zhang, Q. (2006) Effect of conventional and organic production practices on the prevalence and antimicrobial resistance of *Campylobacter* spp. in poultry. *Applied Environmental Microbiology* 72, 3600–3607.

Lumpkins, B.S., Batal, A.B. and Lee, M. (2008) The effect of gender on the bacterial community in the gastrointestinal tract of broilers. *Poultry Science* 87, 964–967.

Luoma, A., Markazi, A., Shanmugasundaram, R., Murugesan, G.R., Mohnl, M. and Selvaraj, R. (2017) Effect of synbiotic supplementation on layer production and cecal *Salmonella* load during a *Salmonella* challenge. *Poultry Science* 96, 4208–4216.

Malik, J.K., Prakash, A., Srivastava, A.K. and Gupta, R.C. (2019) Synbiotics in animal health and production. In *Nutraceuticals in Veterinary Medicine*. Springer, Cham, pp. 287–301.

Marchesi, J.R. and Ravel, J. (2015) The vocabulary of microbiome research: a proposal. *Microbiome* 3, 1–3.

MAP (2021a) *Market Analysis Report; Prebiotics Market Size, Share & Trends Analysis Report By Ingredients (FOS, Inulin, GOS, MOS), By Application (Food & Beverages, Dietary Supplements, Animal Feed), By Region, And Segment Forecasts, 2022–2030,* 1–80.

MAP (2021b) *Market Analysis Report; Probiotics Market Size, Share & Trends Analysis Report By Product (Probiotic Food & Beverages, Probiotic Dietary Supplements), By Ingredient (Bacteria, Yeast), By End Use, By Distribution Channel, And Segment Forecasts, 2021–2030,* 1–135.

Mavri, A. and Smole Možina, S. (2013) Development of antimicrobial resistance in *Campylobacter jejuni* and *Campylobacter coli* adapted to biocides. *International Journal of Food Microbiology* 160, 304–312.

Meade, K.G., Narciandi, F., Cahalane, S., Reiman, C., Allan, B. and O'Farrelly, C. (2009) Comparative *in vivo* infection models yield insights on early host immune response to *Campylobacter* in chickens. *Immunogenetics* 61, 101–110.

Medvecky, M., Cejkova, D., Polansky, O., Karasova, D., Kubasova, T. *et al.* (2018) Whole genome sequencing and function prediction of 133 gut anaerobes isolated from chicken caecum in pure cultures. *BMC Genomics* 19, 1–15.

Mohan, V. (2015) The role of probiotics in the inhibition of *Campylobacter jejuni* colonization and virulence attenuation. *European Journal of Clinical Microbiology and Infectious Diseases* 34, 1503–1513.

Molina, F., Simancas, A., Ramírez, M., Tabla, R., Roa, I. and Rebollo, J.E. (2021) A new pipeline for designing phage cocktails based on phage-bacteria infection networks. *Frontiers in Microbiology* 12, 149.

Monod, J. (1949) The growth of bacterial cultures. *Annual Reviews in Microbiology* 3, 371–394.

Mor-Mur, M. and Yuste, J. (2010) Emerging bacterial pathogens in meat and poultry: An overview. *Food and Bioprocess Technology* 3, 24–35.

Mortada, M., Cosby, D.E., Shanmugasundaram, R. and Selvaraj, R.K. (2020) *In vivo* and *in vitro* assessment of commercial probiotic and organic acid feed additives in broilers challenged with *Campylobacter coli*. *Journal of Applied Poultry Research* 29, 435–446.

Nguyen, T.T.T., Allan, B., Wheler, C., Köster, W., Gerdts, V. and Dar, A. (2021) Avian antimicrobial peptides: *in vitro* and in ovo characterization and protection from early chick mortality caused by yolk sac infection. *Scientific Reports* 11, 1–14.

Nhung, N.T., Chansiripornchai, N. and Carrique-Mas, J.J. (2017) Antimicrobial resistance in bacterial poultry pathogens: A review. *Frontiers in Veterinary Science* 4, 126.

Nordentoft, S., Mølbak, L., Bjerrum, L., de Vylder, J., van Immerseel, F. and Pedersen, K. (2011) The influence of the cage system and colonisation of *Salmonella* Enteritidis on the microbial gut flora of laying hens studied by T-RFLP and 454 pyrosequencing. *BMC Microbiology* 11, 1–11.

O'Neill, J. (2016) Tackling drug-resistant infections globally: final report and recommendations. Wellcome Trust and HM Government. Available at: https://amr-review.org/sites/default/files/160525_Final%20paper_with%20cover.pdf (accessed 6 January 2023).

Osman, K.M., Kappell, A.D., Elhadidy, M., Elmougy, F., El-Ghany, W.A.A. *et al.* (2018) Poultry hatcheries as potential reservoirs for antimicrobial-resistant *Escherichia coli*: A risk to public health and food safety. *Scientific Reports* 8, 5859.

Pan, X.-D., Chen, F.-Q., Wu, T.-X., Tang, H.-G. and Zhao, Z.-Y. (2009) Prebiotic oligosaccharides change the concentrations of short-chain fatty acids and the microbial population of mouse bowel. *Journal of Zhejiang University Science B* 10, 258–263.

Pandey, K.R., Naik, S.R. and Vakil, B.V. (2015) Probiotics, prebiotics and synbiotics - a review. *Journal of Food Science and Technology* 52, 7577–7587.

Pires, D.P., Oliveira, H., Melo, L.D.R., Sillankorva, S. and Azeredo, J. (2016) Bacteriophage-encoded depolymerases: their diversity and biotechnological applications. *Applied Microbiology and Biotechnology* 100, 2141–2151.

Rahimi, S., Kathariou, S., Fletcher, O. and Grimes, J.L. (2019) Effect of a direct-fed microbial and prebiotic on performance and intestinal histomorphology of turkey poults challenged with *Salmonella* and *Campylobacter*. *Poultry Science* 98, 6572–6578.

Rehman, H.U., Vahjen, W., Awad, W.A. and Zentek, J. (2007) Indigenous bacteria and bacterial metabolic products in the gastrointestinal tract of broiler chickens. *Archives of Animal Nutrition* 61, 319–335.

Richards, P.J., Connerton, P.L. and Connerton, I.F. (2019) Phage biocontrol of campylobacter jejuni in chickens does not produce collateral effects on the gut microbiota. *Frontiers in Microbiology* 10, 476.

Ricke, S.C. (2003) Perspectives on the use of organic acids and short chain fatty acids as antimicrobials. *Poultry Science* 82, 632–639.

Roberfroid, M., Gibson, G.R., Hoyles, L., McCartney, A.L., Rastall, R. *et al.* (2010) Prebiotic effects: metabolic and health benefits. *British Journal of Nutrition* 104, S1–S63.

Robinson, K., Becker, S., Xiao, Y., Lyu, W., Yang, Q. *et al.* (2019) Differential Impact of Subtherapeutic Antibiotics and Ionophores on Intestinal Microbiota of Broilers. *Microorganisms* 7, 282.

Rostami, F., Taherpour, K., Ghasemi, H.A., Akbari Gharaei, M. and Shirzadi, H. (2022) Efficacy of Scrophularia striata hydroalcoholic extract and mannan-oligosaccharide on productive performance, intestinal bacterial community, and immunity in broiler chickens after infection with *Campylobacter jejuni*. *Animal Feed Science and Technology* 285, 115–217

Rubio, L.A., Peinado, M.J., Ruiz, R., Suárez-Pereira, E., Ortiz Mellet, C. and García Fernández, J.M. (2015) Correlations between changes in intestinal microbiota composition and performance parameters in broiler chickens. *Journal of Animal Physiology and Animal Nutrition* 99, 418–423.

Saint-Cyr, M.J., Guyard-Nicodème, M., Messaoudi, S., Chemaly, M., Cappelier, J.M. *et al.* (2016) Recent advances in screening of anti-*Campylobacter* activity in probiotics for use in poultry. *Frontiers in Microbiology* 7, 553.

Sánchez, A.M., Bennett, G.N. and San, K.Y. (2006) Batch culture characterization and metabolic flux analysis of succinate-producing *Escherichia coli* strains. *Metabolic Engineering* 8, 209–226.

Sandilands, V., Whyte, F., Williams, L.K., Wilkinson, T.S., Sparks, N.H.C. and Humphrey, T.J. (2018) Reliably colonising broiler chickens with *Campylobacter* spp. using a litter-based method. *British Poultry Science* 59, 698–702.

Santini, C., Baffoni, L., Gaggia, F., Granata, M., Gasbarri, R. *et al.* (2010) Characterization of probiotic strains: An application as feed additives in poultry against *Campylobacter jejuni*. *International Journal of Food Microbiology* 141, S98–S108.

Sasaki, Y., Maruyama, N., Zou, B., Haruna, M., Kusukawa, M. *et al.* (2013) *Campylobacter* Cross-Contamination of Chicken Products at an Abattoir. *Zoonoses Public Health* 60, 134–140.

Schmelcher, M. and Loessner, M.J. (2021) Bacteriophage endolysins – extending their application to tissues and the bloodstream. *Current Opinions in Biotechnology* 68, 51–59.

Shapiro, S.K. and Sarles, W.B. (1949) Microorganisms in the Intestinal Tract of Normal Chickens. *Bacteriology* 58, 531–544.

Smialek, M., Burchardt, S. and Koncicki, A. (2018) The influence of probiotic supplementation in broiler chickens on population and carcass contamination with *Campylobacter* spp. – Field study. *Research in Veterinary Science* 118, 312–316.

Sørensen, T.L., Blom, M., Monnet, D.L., Frimodt-Møller, N., Poulsen, R.L. and Espersen, F. (2001) Transient intestinal carriage after ingestion of antibiotic-resistant *Enterococcus faecium* from chicken and pork. *New England Journal of Medicine* 345, 1161–1166.

Stanley, D., Hughes, R.J. and Moore, R.J. (2014) Microbiota of the chicken gastrointestinal tract: Influence on health, productivity and disease. *Applied Microbiology and Biotechnology* 98, 4301–4310.

Stern, N.J., Svetoch, E.A., Eruslanov, B.V., Perelygin, V.V., Mitsevich, E.V. *et al.* (2006) Isolation of a *Lactobacillus salivarius* strain and purification of its bacteriocin, which is inhibitory to *Campylobacter jejuni* in the chicken gastrointestinal system. *Antimicrobial Agents and Chemotherapy* 50, 3111–3116.

Stern, N.J., Eruslanov, B. v., Pokhilenko, V.D., Kovalev, Y.N., Volodina, L.L. *et al.* (2008) Bacteriocins reduce *Campylobacter jejuni* colonization while bacteria producing bacteriocins are ineffective. *Microbial Ecology in Health and Disease* 20, 74–79.

Svetoch, E.A. and Stern, N.J. (2010) Bacteriocins to control *Campylobacter* spp. in poultry – a review. *Poultry Science* 89, 1763–1768.

Tagliaferri, T.L., Jansen, M. and Horz, H.P. (2019) Fighting pathogenic bacteria on two fronts: phages and antibiotics as combined strategy. *Frontiers in Cellular and Infection Microbiology* 9, 22.

Taha-Abdelaziz, K., Astill, J., Kulkarni, R.R., Read, L.R., Najarian, A. *et al.* (2019) *In vitro* assessment of immunomodulatory and anti-*Campylobacter* activities of probiotic *Lactobacilli*. *Scientific Reports* 9, 1–15.

Tanner, S.A., Berner, A.Z., Rigozzi, E., Grattepanche, F., Chassard, C. and Lacroix, C. (2014) *In vitro* continuous fermentation model (PolyFermS) of the swine proximal colon for simultaneous testing on the same gut microbiota. *PLoS One* 9, e94123.

Tiwari, G., Tiwari, R., Pandey, S. and Pandey, P. (2012) Promising future of probiotics for human health: Current scenario. *Chronicles of Young Scientists* 3, 17–28.

Torok, V.A., Ophel-Keller, K., Loo, M. and Hughes, R.J. (2008) Application of methods for identifying broiler chicken gut bacterial species linked with increased energy metabolism. *Applied Environmental Microbiology* 74, 783–791.

Torok, V.A., Hughes, R.J., Ophel-Keller, K., Ali, M. and MacAlpine, R. (2009) Influence of different litter materials on cecal microbiota colonization in broiler chickens. *Poultry Science* 88, 2474–2481.

van Dijk, A., Veldhuizen, E.J.A. and Haagsman, H.P. (2008) Avian defensins. *Veterinary Immunology and Immunopathology* 124, 1–18.

Videnska, P., Sedlar, K., Lukac, M., Faldynova, M., Gerzova, L. *et al.* (2014) Succession and replacement of bacterial populations in the caecum of egg laying hens over their whole life. *PLoS One* 9, e115142.

Videnska, P., Sisak, F., Havlickova, H., Faldynova, M. and Rychlik, I. (2013) Influence of *Salmonella enterica serovar* Enteritidis infection on the composition of chicken cecal microbiota. *BMC Veterinary Research* 9, 1–8.

Waite, D.W. and Taylor, M.W. (2015) Exploring the avian gut microbiota: Current trends and future directions. *Frontiers in Microbiology* 6, 673.

Wang, Y., Wang, M., Shan, A. and Feng, X. (2020) Avian host defense cathelicidins: structure, expression, biological functions, and potential therapeutic applications. *Poultry Science* 99, 6434–6445.

WHO (2015) World Health Organization: WHO estimates of the global burden of foodborne diseases. Available at: https://www.who.int/publications/i/item/9789241565165 (accessed 6 January 2023).

WHO (2022) World Health Organization: Draft WHO Global Strategy for Food Safety 2022-2030. Available at: https://www.who.int/publications/i/item/9789240057685 (accessed 6 January 2023).

Wine, E., Gareau, M.G., Johnson-Henry, K. and Sherman, P.M. (2009) Strain-specific probiotic (*Lactobacillus helveticus*) inhibition of *Campylobacter jejuni* invasion of human intestinal epithelial cells. *FEMS Microbiology Letters* 300, 146–152.

Yeoman, C.J., Chia, N., Jeraldo, P., Sipos, M., Goldenfeld, N.D. and White, B.A. (2012) The microbiome of the chicken gastrointestinal tract. *Animal Health Research Reviews* 13, 89–99.

Zhang, Y., Lin, Y., Galgano, S., Houdijk, J., Xie, W. *et al.* (2022) Recent progress in phage therapy to modulate multidrug-resistant *Acinetobacter baumannii*, including in human and poultry. *Antibiotics* 11, 1406.

Zhao, L., Wang, G., Siegel, P., He, C., Wang, H. *et al.* (2013) Quantitative genetic background of the host influences gut microbiomes in chickens. *Scientific Reports* 3, 1–6.

Zhou, W., Chen, Y., Roh, T., Lin, Y., Ling, S. *et al.* (2018) Multifunctional bioreactor system for human intestine tissues. *ACS Biomaterials Science and Engineering* 4, 231–239.

Zimmermann, B., Bauer, E. and Mosenthin, R. (2001) Pro and prebiotics in pig nutrition – potential modulators of gut health? *Journal of Animal Feed Science* 10, 47–56.

CHAPTER 6
Is Competitive Exclusion a Useful New Technology?

Ivan Rychlik*
Veterinary Research Institute, Brno, Czech Republic

Abstract

Gallus gallus has evolved to be hatched in nests and live for up to 20 years, whereas broilers in commercial production are hatched in hatcheries and live for approximately 1 month. 'Long-lived' layers live for approximately 1 year. Forgetting these facts results in mistaken conclusions on what is normal, including what is normal gut microbiota. If newly hatched chicks are kept in a contact with adult hens, they acquire an adult type of gut microbiota within first 5 days of their life. Consequently, they become resistant to gut pathogens such as *Salmonella* or *Clostridium perfringens*. Moreover, the transfer of microbiota between hens and their offspring is not restricted to gut microbiota but also includes skin or respiratory tract microbiota. Laboratory experiments with broilers from hatcheries have led to the mistaken conclusion of another specific type of microbiota, i.e. the caecal mucosal microbiota. Unlike the luminal microbiota, dominated by *Bacteroidetes* and *Firmicutes*, the mucosal microbiota of chicken caecum is dominated by *Helicobacter*, *Mucispirillum*, *Desulfovibrio* and *Spirochaetes*. Targets for competitive exclusion need not be sought only within chickens themselves but also within their environment. One of the possible targets for biological intervention might be hatcheries in which we begin to observe negative correlations between the presence of particular bacterial species (*Flavobacteria* and *Sphingobacteria*) and the presence of *Salmonella*, or a positive correlation with flock performance after placement at farms. Understanding interactions in complex bacterial population within chicken production therefore creates an opportunity for the targeted competitive exclusion of undesired bacterial species.

*Corresponding author: ivan.rychlik@vri.cz

DOI: 10.1079/9781800622746.0006

6.1 Introduction

Chickens represent the most widespread source of animal protein, acceptable for nearly all communities worldwide, and the highly efficient commercial production of chickens is a daily practice in all continents. However, this has led to the situation when we have forgotten several basic facts about *Gallus gallus* and mistakenly consider commercial standards as a reference point. For example, chickens evolved to be hatched in nests and to be in close contact with their parents from the very first moments of life. Instead, commercial chickens are hatched in the clean environment of hatcheries without any contact with adult birds. Next, we consider broilers of 35 days of age as a final product and egg layers over 1 year of age as old and exhausted by egg production. However, *G. gallus* can live up to 20 years and reach sexual maturity around week 18 of life. Any discussion on stabilization of the immune system or gut microbiota in broilers is highly misleading and a consequence of human ignorance of the basic biological facts of *G. gallus*. The following chapter therefore aims to discuss these discrepancies and possible interventions that may at least partially correct the conditions prevalent in commercial production. Proposed interventions could improve the living condition of these chickens whilst otherwise preserving current production standards. Discussed interventions will focus on the microbial colonization of chickens and the microbial composition of the chicken environment.

6.2 Competitive Exclusion in the Chicken Caecum

It is well known that oral inoculation of newly hatched chicks from hatcheries with faecal material from adult hens results in an immediate increase of their resistance to *Salmonella* and other enteric pathogens (Nurmi and Rantala, 1973; Rantala and Nurmi, 1973). This is associated with the rapid colonization of the chicken intestinal tract with adult-type microbiota, which suppresses *Salmonella* colonization. This experiment can be set up by co-housing a randomly selected adult egg-laying hen (around 40 weeks of age) with chicks hatched in hatcheries (Kubasova *et al.*, 2019). This co-housing can be as short as for the first 24 hours of chick's life. This time is sufficient to ensure natural transfer of gut microbiota from hens to chicks and to increase chicken resistance to *Salmonella* challenge. A comparison of the caecal microbiota of one-week-old chicks raised in the absence of adult hens with those kept in a contact with an adult hen allows the identification of gut microbiota members that are transferred from hens to offspring. These are of course absent from gut microbiota of chicks from hatcheries raised in the absence of any hen, that is. under the conditions in commercial production. Moreover, the microbiota transferred from hens to chicks can be predicted to be responsible for the competitive exclusion of *Salmonella* and other enteric pathogens from the chicken intestinal tract (Nurmi and Rantala, 1973; Kubasova *et al.*, 2019). Such comparisons convincingly showed that strict bacterial species transferred from hens to chicks originate from the class *Negativicutes* (phylum *Firmicutes*) and from phyla *Bacteroidetes* and *Proteobacteria* (Fig. 6.1).

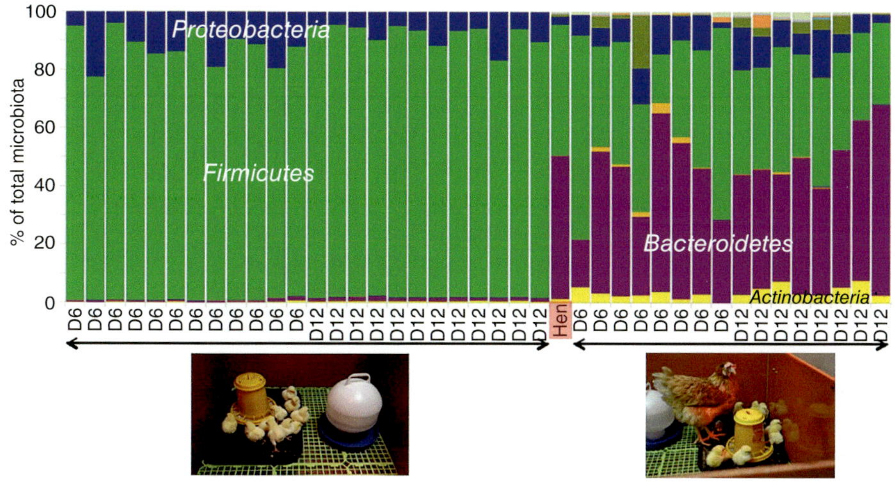

Fig. 6.1. The composition of caecal microbiota in one-week-old chicks with or without contact with an adult hen. Each column represents the caecal microbiota of a single chick. D6 and D12 represents the age of chicks in days. Contact with the hen lasted for 24 hours only before the chicks were moved to a new clean box.

Interestingly, if the donor hen is positive for *Campylobacter*, this is also among the species that are efficiently transferred to contact chicks. Moreover, *Campylobacter* usually forms around 0.1% of the total caecal microbiota in adult hens, whereas it can form around 5% of the caecal microbiota in one-week-old contact chicks (Polansky *et al.*, 2015). Adult hens can therefore control the multiplication of *Campylobacter* while newly hatched chicks are highly susceptible, especially when co-inoculated with the complex microbiota of adult hens. The use of adult hens as donors of microbiota in broilers, for example, by adding one hen per 1000 chicks, may potentially increase chick resistance to *Salmonella*; however, it may also introduce a new problem with *Campylobacter*. When the increased susceptibility of the chicks is known for *Campylobacter*, it is possible to extrapolate that the same may occur with other opportunistic pathogens, of bacterial and other sources.

6.3 Why are Some Bacterial Taxa Easily Transferred from Hens to Chickens?

The gut anaerobes that are somewhat resistant to air oxygen are usually those that are not transferred from hens to offspring, although this is not experimentally proven yet (Kubasova *et al.*, 2021). These include all spore-forming bacteria mostly from the order Clostridiales, aerotolerant bacterial species such as *Lactobacilli* or facultative anaerobes like *Escherichia coli*. Their presence in the intestinal tract of chickens is a consequence of the permanent exposure of chicks to spores or viable cells present in the environment and their temporarial multiplication in the intestinal tract before they are excreted back to the environment by peristalsis. Their ubiquitous presence in the environment and continual chick exposure resulted in these bacteria not evolving mechanisms that would allow them to permanently colonize at the very first opportunity. On the other hand, bacteria that do not exhibit any form of aerobic resistance must use every opportunity for long-term colonization of a new host. Such bacterial species are therefore of special interest for a design of novel probiotics because a single-dose administration is enough for their long-term colonization. If bacterial species from among those not permanently colonizing the chicken intestinal tract are used, their natural ecology should be considered and these need to be administered for prolonged period of time via feed or drinking water.

6.4 Age-dependent Microbiota Development in Chickens

We and others have reported that there is an age-dependent development of the chicken gut microbiota (Stanley *et al.*, 2013; Videnska *et al.*, 2014; Gao *et al.*, 2017; Siegerstetter *et al.*, 2018; Xi *et al.*, 2019). *E. coli* is commonly reported as the first gut colonizer. If the gut microbiota is determined in the caecum of chicks from hatcheries, *E. coli* indeed dominates during the first week of life. Starting from week two, dominance of *E. coli* is suppressed and replaced by bacterial species from the families *Lachnospiraceae*, *Ruminococcaceae* and *Lactobacillaceae*. Different species from the phylum *Bacteroidetes* usually appear after these. Later gut colonizers also include bacterial species from the class *Negativicutes* (*Megamonas*, *Megasphaera*, *Veillonella*, *Dialister* or *Phascolarctobacterium*) and the strictly anaerobic *Proteobacteria* (*Sutterella*, *Parasutterella*, *Succinatimonas* or *Anaerobiospirillum*).

On the basis of these observations, age-dependent development of gut microbiota has been proposed, including noting the necessary 'anaerobization' of the intestinal tract by facultative anaerobes to make the environment permissive for strict anaerobes. However, the experiments with chicks and contact hens show that this way of thinking is mostly misleading. Studies reporting age-dependent development of gut microbiota in broilers are correct in their observations but mistaken in their description of real biology. The reported age-dependent development in chicks-alone studies strengthens the role of parents as a source of correct microbiota. If hens are present, the development of gut microbiota is a matter of the first few days of life, definitively less than one week. This does not exclude subsequent shaping of gut microbiota composition depending on diet composition. We also know that some bacterial species from the gut microbiota grow faster (*Megamonas hypermegale*, *Bacteroides caecicola*) while others (*Bacteroide gallinaceum*, *Marseilla massiliensis*) are slower, demanding a few more days to establish themselves in gut microbiota. However, reports on the age-dependent development of gut microbiota are only a consequence of using chicks from hatcheries and raising them on their own, without access to an adult type of avian gut microbiota.

Attempts to define an age when the gut microbiota of commercially raised chickens stabilizes, or which species appears earlier and which later, therefore do not consider the natural ecology and ethology of these birds. These measures are therefore confounding, biologically incorrect and result from ignoring the basic facts. Viewing the same situation from a different angle, using caecal samples from one-week-old broilers for the culture of gut microbiota adapted to the conditions is a mistaken approach. Due to the absence of microbiota from adult hens, microbiota present in the caecum of one-week-old broilers represents the environmental microbiota that is also capable of replication in the chicken intestine. Since spore formers and aerotolerant bacteria survive in the environment, these are usually the first ones reported in the intestinal tract. However, these species are far from generally considered to be adapted to young chickens because if there were a source, i.e., the hen, the gut microbiota of one-week-old chicks would be different.

6.5 Microbiota–Host Adaptation

Gut microbiota of omnivorous warm-blooded animals, including humans, consists of bacterial species belonging to four main phyla: *Bacteroidetes*, *Firmicutes*, *Proteobacteria* and *Actinobacteria*. Nearly every publication on the gut microbiota of chickens, mice, pigs or humans starts with such information (Videnska *et al.*, 2014; Li *et al.*, 2016; Kubasova *et al.*, 2017; Li *et al.*, 2018). However, does this mean that the gut microbiota of these animal species is formed by the same bacterial species? To address this question, we compared samples collected from chickens, pigs and humans (Karasova *et al.*, 2022). Interestingly, there were species that exhibited high host adaptation, as well as species that could be found in all three hosts. *E. coli* was the species without any host preference. Using *E. coli* as an example, this of course does not necessarily mean that the same clones of *E. coli* are found in humans, pigs and chickens. It is likely that at strain, clone or isolate level, host preference of particular clones will also be recorded in *E. coli*. The same is true for spore-forming *Clostridiales* since the same species can be found in gut microbiota of chickens, pigs and humans. In contrast to host adaptation of *E. coli* or *Clostridiales* at clone level, in *Bacteroides* or *Megasphaera*, host adaptation can be seen at species level, i.e. at higher taxonomic level. The same genera are shared by chickens, pigs and humans; chickens are colonized by *Megasphaera stantonii*, whereas *Megasphaera elsdenii* is characteristic for pigs and humans. Chickens are colonized by *Bacteroides caecicola*, *B. caecigallinarum*, *B. plebeius*, *B. barnesiae* or *B. mediterraneensis*, whereas humans are colonized by *Bacteroides dorei*, *B. ovatus*, *B. uniformis* or *B. thetaiotaomicron* (Kollarcikova *et al.*, 2020). This is difficult to prove experimentally but it is very likely this phenomenon is affected by the survival ability of bacteria in an aerobic environment. Taxa that exhibit some form of aerobic resistance can persist in the environment for a prolonged period and from this environment they can come into contact with humans, pigs or chickens. Oxygen-resistant taxa can spread over great physical distances and, after being released into the environment by pigs, they can persist in the form of endospores and can be then spread by air dust to poultry barns. Conversely, in gut anaerobes that do not exhibit any form of aerobic existence and are inactivated by oxygen within hours of exposure, the likelihood of their spread over great distances is limited. They can therefore be spread only over short distances, by contact among individuals of the same herd or flock (Karasova *et al.*, 2022). Consequently, such separation is the first step to speciation.

6.6 Mucosal Microbiota in the Chicken Caecum

It is not difficult to imagine that microbiota composition differs in the gut lumen and the gut mucus as these two sites differ considerably from a nutritional standpoint. The gut lumen is filled by digesta of feed origin, while mucus is formed mostly by host polysaccharides and proteins. Although there have been general discussions on this topic, few papers on gut mucosal microbiota have been published. When we tested this idea several years ago, perhaps surprisingly, we did not find any clear difference between mucosal and luminal microbiota in one-week-old chickens. However, the problem in this case was the age and origin of these chickens. These chicks originated from a commercial hatchery and did not come in contact with adult birds. As the luminal bacteria are transferred from hens to offspring, these chicks did not get the chance to be inoculated and then to develop their mucosal microbiota. When we later tested the mucosal microbiota in the caeca from adult hens, we detected several taxa that are usually referred to as minority taxa in studies analysing the microbiota of faecal or caecal samples. When mucosal samples from adult hens are collected, the mucosal microbiota of chicken caecum does not consist of *Bacteroidetes*, *Firmicutes*, *Proteobacteria* and *Actinobacteria* but of *Mucispirillum* sp. (phylum *Deferribacteria*), *Treponema* sp. and *Brachyspira innocens* (*Spirochaeta*), *Helicobacter pullorum* (previously phylum *Proteobacteria*, currently new phylum *Campylobacterota*) and *Desulfovibrio* sp. (previously phylum *Proteobacteria*, currently new phylum *Desulfobacterota*) (Fig. 6.2). Using metagenomic or whole-genome DNA sequencing we confirmed that *Helicobacter*, *Mucispirillum*, *Desulfovibrio* and *Brachyspira* encode genes for motility and chemotaxis. Since *Treponema* is flagellated, it seems that motility is required for the colonization of the mucus layer. Experiments performed with newly hatched chicks and contact hens showed that the mucosal microbiota can also be efficiently transferred between parents and their offspring and their absence in the intestinal tract of chicks from hatcheries is not caused by immaturity of the intestinal tract of chicks but by the absence of sources. Newly hatched chicks can therefore be colonized by the mucosal microbiota from the very first moments of their life.

The function of mucosal microbiota is not well understood, as these species are not yet fully recognized as mucus associated. Despite this, it is simple to imagine that the mucosal microbiota is in more intimate contact with the host, with its epithelial cells as well as immune cells associated with the epithelium (Herp *et al.*, 2019). On the other hand, luminal bacteria, i.e. *Firmicutes* or *Bacteroidetes*, can communicate with the chicken host only by diffusible by-products of their metabolism. Although the release of acetate, propionate, butyrate or lactate (and many other molecules) is beneficial for the host,

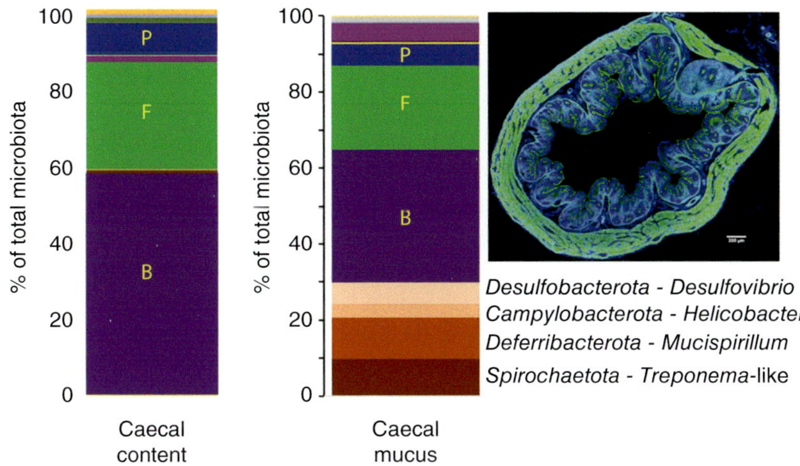

Fig. 6.2. Mucosal microbiota. The luminal caecal microbiota of adult hens is formed by the phyla *Bacteroidetes* (B), *Firmicutes* (F) and *Proteobacteria* (P). Mucosal surfaces of the same hens were colonized by *Treponema*-like *Spirochaeta*, *Mucispirillum*, *Helicobacter* and *Desulfovibrio*. Based on unpublished data from the author's laboratory.

which can utilize such substrates in aerobic metabolism and respiration, this way of communication is clearly different from the contact between the host and mucus-colonizing bacteria, e.g. by flagella stimulating TLR5 receptors. Moreover, mucosal microbiota can produce and release by-products of their anaerobic metabolism. This is evidenced by the fact we have only recorded genes for the butyrate metabolic pathway in the genome of *Brachyspira innocens*, among the few bacterial species colonizing caecal mucus. Simultaneously, one has to be very careful with the role of mucus-associated microbiota because these bacteria might be on a fine border between commensals and pathogens. At low or moderate abundance, they may stimulate the mucosal gut immune system. It can be easily imagined, however, that their overgrowth might result in an inappropriate inflammatory response associated with tissue damage and therefore lead to a decrease in chicken performance. Many more experiments with mucosal microbiota are therefore needed before it is possible to decide on their probiotic potential.

6.7 Competitive Exclusion Outside the Intestinal Tract

Competitive exclusion need not be restricted to the intestinal tract. In experiments with hens and contact chicks, we observed differential development of skin or tracheal microbiota in contact and control chickens (unpublished results). Skin and tracheal microbiota were similar to each other but differed in control chicks and in contact chicks kept with an adult hen. Tracheal and skin microbiota of control one-week-old chicks were dominated by spore-forming or aerobic microbiota like *Clostridium disporicum*, *Blautia argi*, *[Ruminococcus] torques* (spore formers) or *Flavobacterium succinicans*, *Cutibacterium acnes* and *E. coli* (aerobic microbiota). Chicks in contact with adult hens showed an increased abundance of *Lactobacilli* both in tracheal and skin microbiota. We also noticed that *Lactobacillus gallinarum* and *L. aviarius* usually dominated in the small intestine, while *L. salivarius*, *L. ingluviei*, *L. reuteri* and *L. johnsonii* are more frequently detected on skin and in the trachea and crop. It is therefore possible that the administration of probiotic *Lactobacilli* intended for intestinal tract colonization in fact results in the colonization of the respiratory tract and improves chicken performance in this way. Continuous administration of *Lactobacilli* via drinking water, especially during the first week of life, may result in the colonization of the skin and trachea, as would happen if the chicks were in contact with an adult hen. This may lead to *Lactobacilli* release into the environment, particularly in litter (see Section 6.11), where it may suppress the growth of opportunistic pathogens.

6.8 Microbiota and Competitive Exclusion Outside the Chicken Host

Chickens are surrounded by an environment that is not sterile. Instead, it is populated by diverse microbial communities. Chicks are exposed to environmental bacterial flora immediately after hatching in hatcheries. Remains of eggshells and egg contents represent material rich in biological molecules and substrates that can be utilized by bacteria for their growth. In addition, temperature and moisture inside hatching incubators support bacterial growth. Bacteria from the eggshell waste therefore represent the first antigen to which newly hatched chicks are exposed. After transport to farms, chicks are exposed to microbiota in bedding, drinking water and feed and, during fattening, faecal material is released into bedding, forming litter with its specific microbiota. At each step, particularly in the eggshell waste in the hatcheries and in the litter, bacterial microbiota is rich in a qualitative as well as quantitative manner, and therefore suitable for active interventions, with the aim to improve total performance. There is therefore no reason not to adapt the competitive exclusion concept for biological cleaning and to utilize microbe–microbe interactions outside the chicken host.

6.9 Eggshell Waste Microbiota

In a number of eggshell waste samples collected from the floor of hatcheries after removal of the chicks, we observed quite complex microbiota. The most frequent taxa included *Acinetobacter*, *E. coli*, *Chryseobacterium*, *Clostridium disporicum*, *Comamonas*, *Enterococcus*, *Proteus*, *Pseudomonas*, *Sphingobacterium* and *Brevibacterium*. *Lactobacilli* were quite rare in this type of sample, forming less than 0.01 % of total microbiota. We also observed that when *Acinetobacter* and other aerobic *Proteobacteria* dominated, environmental *Bacteroidetes* (*Flavobacteria* and *Sphingobacteria*) were suppressed, and vice versa, when *Flavobacteria* and *Sphingobacteria* dominated, *Proteobacteria* including *Salmonella* were suppressed (unpublished results). It is possible that the dominance of *Proteobacteria* or *Flavobacteria* is due to the physical condition of the material (wet or drier, etc.). However, direct antagonistic interactions between these two groups of bacteria cannot be excluded. These interactions require more study, especially as we recorded a minor positive correlation between the higher abundance of *Flavobacteria* and *Sphingobacteria* in the eggshell waste and lower fatalities in flocks during the first week after placement at farms and vice versa (positive correlation was observed between higher *Enterobacteria* abundance in eggshell waste and higher fatality rates in chicks during first week after placement at farms). It might be possible to identify bacteria that suppress the multiplication of *E. coli* and *Salmonella* in the eggshell waste and decrease fatalities of chicks after their placement on farms and use these as probiotics or biological cleaning products to improve zoo-hygienic conditions in hatcheries and at farms.

6.10 Drinking Water Microbiota

Microbiota in drinking water in chicken farms have not been well characterized, even though water consumption is continual during production, and thus inoculation with microbiota from water is permanent. Since it is well known that overgrowth of *Lactobacilli* results in a decrease in pH, down to values that suppress growth of competing microbiota, a continuous supplementation of drinking water with low levels of *Lactobacillus* species characteristic for chickens can be considered, based upon their intrinsic ability to replicate in the intestine and the environment. Such *Lactobacilli* can not only protect the water supply but will then be spread into the farm environment and may colonize the chicken respiratory tract, skin, crop and litter.

6.11 Litter Microbiota

Litter is the last type of sample that can be considered for targeted competitive exclusion. The definition of the major members of litter microbiota is complicated by the fact that, as litter develops, the microbiota composition changes accordingly. Dry straw bedding is quite poorly populated with bacteria. However, within the first two days after chicken placement at farms, *E. coli* of chicken origin appears in the forming litter. Its persistence in the litter enables *E. coli* recycling in chicks by an oro-faecal route. Dominance of *E. coli* in the litter lasts for the first two weeks after chicken placement, when it is replaced by *Corynebacteria*, saprophytic *Staphylococci* and *Lactobacilli*. During the first month, the major representatives of litter microbiota include *Staphylococcus saprophyticus*, *Weissella jogaejeotgali*, *Corynebacterium casei*, *E. coli*, *Enterococcus hirae*, *L. johnsonii*, *L. salivarius*, *L. reuteri* and *Lactobacillus gallinarum*. Of these, *E. coli*, *Enterococcus* and *Lactobacilli* are common also to the chicken intestinal tract and their presence in the litter is probably a consequence of chicken placement and the release of faecal material. On the other hand, *S. saprophyticus*, *Weissella jogaejeotgali* and *Corynebacterium casei* are not reported as chicken (gut) microbiota and might be considered as litter-specific bacterial species. During month two, *E. coli* and *Enterococcus* are usually suppressed in their abundance and *Lactobacilli* still prevail. Newly appearing taxa include *Brachybacterium*, *Brevibacterium*, *Jeotgalicoccus* and *Mammaliicoccus* (*Staphylococcus*) *lentus*. As litter develops further, *Lactobacilli* becomes a minority taxa and bacteria commonly associated with environments of high osmolarity begin to dominate. These include genera *Salinicoccus*, *Sphingobacterium*, *Geofilum*, *Oceanobacillus*, *Dietzia* or *Yaniella* (Cressman *et al.*, 2010; De Cesare *et al.*, 2019; Wang *et al.*, 2016; Johnson *et al.*, 2018).

Similar to eggshell waste microbiota, it is not known whether such development is determined by the physico-chemical conditions in the forming litter or whether this development can be potentiated by active inoculation of the bedding with particular taxa. Nevertheless, inoculation of the bedding with *Corynebacteria*, saprophytic *Staphylococci* and *Lactobacilli* just before chicken placement at farms may represent an interesting intervention, since these bacteria are naturally present in the bedding for the first two months, i.e. within the time of broiler production. The fact that these bacteria are present in the litter should be considered but, at the same time, need not be used as justification that there are natural environmental sources of these bacteria. It might be useful to inoculate the bedding as early as possible, to speed up proper colonization. This will result in a microbiologically reproducible environment as early as possible, for example, an environment suppressing *E. coli* or *Enterococcus hirae* multiplication from the first moments of their release by chickens. Each batch of chickens will thus be placed in the same environment with a secured supply of desired microbiota.

6.12 The Most Widespread Microbiota in Chicken Production

A comparison of chicken gut, skin or tracheal microbiota with microbiota in the environment allows the definition of the most common bacterial taxon in chicken production, *E. coli*, which can be found in all internal areas of chickens including the strictly anaerobic caecum and it is also common in environmental samples. Although not the most abundant in absolute counts, *E. coli* usually dominates in the external environment where it can use its aerobic metabolism. *E. coli* can form up to 10% of total bacterial flora in eggshell waste, drinking water or litter, but it forms only around 0.1% of the total caecal microbiota, suggesting that the presence of *E. coli* in the chicken gut is forced by its ubiquitous presence in the environment. Despite its common presence, *E. coli* is not a typical bacterium adapted to the gut environment. There are many species that are better adapted to the chicken intestinal tract than *E. coli*.

The second group of bacteria found in samples from chicken production and its environment includes different *Lactobacillus* species. *Lactobacilli* are found in the skin, trachea, crop, small intestine, caecum or in litter. The origin of *Lactobacilli* is environmental but *Lactobacilli* do not originate from hatcheries. Once the chickens become colonized by *Lactobacilli*, continuous voiding of mostly small-intestine digesta rich in *Lactobacilli* secures their presence in environmental samples like litter, drinking water or feed.

The last group of broadly distributed species in chicken production includes spore-forming *Clostridium disporicum* and *Ruminococcus torque*. Their broad distribution suggests that these species must be both efficient spore formers as well as spore germinators. The broad association of two spore-forming species with chickens may explain the positive effects *Bacillus* spores have on chicken gut health. *Bacillus* spores can mimic and replace spores of *C. disporicum* and *R. torque*. Since bacterial endospores are highly specific for prokaryotes, it is tempting to speculate that some of the structures and molecules on the spore surface are sensed by the chicken immune system with appropriate responses resulting in activation of innate immunity and increased resistance to enteric pathogens.

6.13 Conclusion

The characterization of microbial populations by DNA sequencing allows a definition of their composition with little bias. The characterization of the microbiota composition is a necessary first step but should not be the only measure when dealing with microbial populations. Studies focused on the culture of bacterial isolates from different type of samples followed by the reintroduction of obtained isolates to their original niche should follow. Only in this way we can fully utilize information on microbial composition in the caecum, trachea, eggshells or litter for active manipulation with microbial communities, to improve conditions across the whole poultry sector. We mentioned in this chapter some of the areas where we can see a potential for future development. Specific microbiota may be selected to treat and repopulate eggshells in hatcheries after egg cleaning and disinfection. Such products of biological cleaning might be applied immediately after cleaning or just prior to the beginning of hatching, on day 19 or 20 of embryonic development. The target in this case would not be the chick, but the environment of hatcheries and the egg surface. When hatching is completed, the already hatched chicks can be treated with chicken-specific microbiota that is typically absent in chicks from hatcheries. An additional step might be colonization of the skin, respiratory tract and bedding. This can be achieved by *Lactobacilli* administration via drinking water. The dosage need not be too high to reduce the cost, since *Lactobacilli* should be able to replicate once they reach the target place, i.e. the skin, trachea, crop or small intestine. When subsequently released into environment and litter, *Lactobacilli* may suppress the multiplication of competing microbiota. Finally, bedding material can be seeded by bacteria characteristic for the litter. The seeding can be done at a low dose because these bacteria will multiply in the matrices.

It may be that some of our predictions will not be viable. However, combined approaches to improve chicken production using probiotics intended directly for chicken colonization combined with microbial products intended for biological cleaning and modification of environment may represent a challenging but interesting way of future development, along with the restoration and further development of the original competitive exclusion concept. If returning to the title of this chapter, 'Is competitive exclusion a useful new technology?', the answer is yes, it is useful and it is more likely to be applied today than it was several years ago. From the evidence presented in this chapter, it is worth highlighting that competitive exclusion should no longer be limited to a gut application but it has the potential to be extended to the respiratory tract, skin and environment.

References

Cressman, M.D., Yu, Z., Nelson, M.C., Moeller, S.J., Lilburn, M.S. and Zerby, H.N. (2010) Interrelations between the microbiotas in the litter and in the intestines of commercial broiler chickens. *Applied and Environmental Microbiology* 76, 6572–6582.

De Cesare, A., Caselli, E., Lucchi, A., Sala, C., Parisi, A. *et al.* (2019) Impact of a probiotic-based cleaning product on the microbiological profile of broiler litters and chicken caeca microbiota. *Poultry Science* 98, 3602–3610.

Gao, P., Ma, C., Sun, Z., Wang, L., Huang, S. *et al.* (2017) Feed-additive probiotics accelerate yet antibiotics delay intestinal microbiota maturation in broiler chicken. *Microbiome* 5, 91.

Herp, S., Brugiroux, S., Garzetti, D., Ring, D., Jochum, L.M. *et al.* (2019) Mucispirillum schaedleri antagonizes Salmonella virulence to protect mice against colitis. *Cell Host and Microbe* 25, 681–694.

Johnson, T.J., Youmans, B.P., Noll, S., Cardona, C., Evans, N.P. *et al.* (2018) A consistent and predictable commercial broiler chicken bacterial microbiota in antibiotic-free production displays strong correlations with performance. *Applied and Environmental Microbiology* 84, e00362-00318.

Karasova, D., Faldynova, M., Matiasovicova, J., Sebkova, A., Crhanova, M. *et al.* (2022) Host species adaptation of obligate gut anaerobes is dependent on their environmental survival. *Microorganisms* 10, 1085.

Kollarcikova, M., Faldynova, M., Matiasovicova, J., Jahodarova, E., Kubasova, T. *et al.* (2020) Different bacteroides species colonise human and chicken intestinal tract. *Microorganisms* 8, 1483

Kubasova, T., Davidova-Gerzova, L., Merlot, E., Medvecky, M., Polansky, O. *et al.* (2017) Housing systems influence gut microbiota composition of sows but not of their piglets. *PLoS One* 12, e0170051.

Kubasova, T., Kollarcikova, M., Crhanova, M., Karasova, D., Cejkova, D. *et al.* (2019) Contact with adult hen affects development of caecal microbiota in newly hatched chicks. *PLoS One* 14: e0212446.

Kubasova, T., Seidlerova, Z. and Rychlik, I. (2021) Ecological adaptations of gut microbiota members and their consequences for use as a new generation of probiotics. *International Journal of Molecular Science* 22, 5471.

Li, S.S., Zhu, A., Benes, V., Costea, P.I., Hercog, R. *et al.* (2016) Durable coexistence of donor and recipient strains after fecal microbiota transplantation. *Science* 352, 586–589.

Li, X., Liang, S., Xia, Z., Qu, J., Liu, H. *et al.* (2018) Establishment of a Macaca fascicularis gut microbiome gene catalog and comparison with the human, pig, and mouse gut microbiomes. *Gigascience* 7, giy100.

Nurmi, E. and Rantala, M. (1973) New aspects of Salmonella infection in broiler production. *Nature* 241, 210–211.

Polansky. O., Sekelova, Z., Faldynova, M., Sebkova, A., Sisak, F. and Rychlik, I. (2015) Important metabolic pathways and biological processes expressed by chicken cecal microbiota. *Applied Environmental Microbiology* 82, 1569–1576.
Rantala, M. and Nurmi, E. (1973) Prevention of the growth of Salmonella infantis in chicks by the flora of the alimentary tract of chickens. *British Poultry Science* 14, 627–630.
Siegerstetter, S.C., Petri, R.M., Magowan, E., Lawlor, P.G., Zebeli, Q. *et al.* (2018) Fecal microbiota transplant from highly feed-efficient donors shows little efect on age-related changes in feed-efficiency-associated fecal microbiota from chickens. *Applied and Environmental Microbiology* 84, e02330-17.
Stanley, D., Geier, M.S., Hughes, R.J., Denman, S.E. and Moore, R.J. (2013) Highly variable microbiota development in the chicken gastrointestinal tract. *PLoS One* 8:e84290.
Videnska, P., Sedlar, K., Lukac, M., Faldynova, M., Gerzova, L. *et al.* (2014) Succession and replacement of bacterial populations in the caecum of egg laying hens over their whole life. *PLoS One* 9, e115142.
Wang, L., Lilburn, M. and Yu, Z. (2016) Intestinal Microbiota of Broiler Chickens As Affected by Litter Management Regimens. *Frontiers in Microbiology* 7, 593.
Xi, Y., Shuling, N., Kunyuan, T., Qiuyang, Z., Hewen, D. *et al.* (2019) Characteristics of the intestinal flora of specific pathogen free chickens with age. *Microbial Pathogenesis* 132, 325–334.

CHAPTER 7
Go With Your Gut – a Scientific Approach to Excellence in Business?

Anne Richmond[1]*, Nicolae Corcionivoschi[2] and Ozan Gundogdu[3]

[1]*Moy Park Ltd, Portadown, Craigavon;* [2]*Agri Food and Biosciences Institute, Belfast;* [3]*London School of Hygiene and Tropical Medicine, London*

Abstract

Dr Anne Richmond has worked in Moy Park for 8 years and has over 30 years' experience working with poultry due to growing up on a poultry farm in Northern Ireland. She has a BSc (Hons) in Biology and PhD in Sustainable Agriculture with her thesis entitled 'An investigation of methane emissions from beef cattle grazing contrasting upland and lowland vegetation types'. Her first role in Moy Park was as a 'Data Analyst' and she introduced JMP statistical software to the business for enhanced data exploration and statistical analysis. Gradually migrating to R&D and formalizing the department in May 2016, she has been involved in many industries leading projects including *Campylobacter* and antibiotic reduction, bird management, welfare and sustainability. She currently co-supervises six PhD students in a range of topics from welfare to novel biopolymers. She has experience of working with the major UK retailers and food service providers and completed time at the Parliamentary Office of Science and Technology as a fellow. In this chapter she will provide an industry perspective and outline Moy Park's journey in the evaluation of probiotics, the challenges posed with conducting research in a commercial environment and the potential for the future, with particular focus on *Campylobacter* reduction. She will also present a summary of research findings within the wider topic of the microbiome, which has been in collaboration with their academic partner.

*Corresponding author: Anne.Richmond@moypark.com

DOI: 10.1079/9781800622746.0007

7.1 Introduction

Conducting applied poultry research in a commercial setting is challenging as there is always the need to maintain operational processes to ensure business integrity and 'balance' in experiment treatments. Moy Park is a unique poultry integration business within the UK as it is responsible for the care of three generations of birds (grandparent, parent and broiler birds). Our farms and farmers are supported by an internal specialist team of nutritionists, feed millers, vets, technical support staff, agricultural engineers and specialists in incubation and hatchery management. There are several benefits to this integration, including sustainability, biosecurity and quality. However, the real strength for our research is that, if we find something that works, we can implement it and it can have a positive influence on the lives of millions of birds.

All research should be underpinned by robust scientific principles and statistical analysis. During the past six years we have also invested in several PhD studentships and have found this to be an excellent way of delivering effective research. It facilitates in the development of subject matter experts, retains skills and experience, and blends new research techniques and thinking with the wealth of experience we have internally.

It is also important for companies to collaborate with academics, contributing to what are often complex research questions. These collaborations lead to peer-reviewed publications. Peer-reviewed science is the core of business philosophy because it gives companies, customers and consumers the confidence that the work is to the highest levels of scientific merit and that it is not just a commercial interpretation. To accomplish this, and to ensure companies remain at the forefront of research and enable access to experts, it is important to have a significant network of academic partners who can be drawn on to address challenges in partnership with industry.

Many times within commercial businesses you will hear the phrase 'let's have a look-see'. This is something we have completely moved away from within our business model for several reasons. The main reason is that these situations still require investment in both time and money and often their results can be misleading, which can result in further costs being incurred as both time and money. When embarking on trial work, it is crucial to define the specific question that you are trying to answer; Albert Einstein famously said 'If I had an hour to solve a problem and my life depended on it, I would use the first fifty-five minutes determining the proper question to ask'. To make it worthwhile investing time, money and effort, we have to ensure a robust experimental design and that we have a question worth answering.

Once a well-defined research question has been formed, the next, crucial stage is to stipulate the type of data required that is relevant within the confinement of what is physically and financially possible. Often, we can become over enthusiastic and keep adding requirements, but this can compromise and dilute what is essential. It is at this stage that the sample size and type of statistical analysis that will be deployed should be agreed, and a Power test considered. Prior to commencing any practical data collection, the statistical analysis that will be used should be known. Collecting large, ad hoc amounts of data with no defined sampling intervals or protocol, with the hope of extracting an answer at a later point, does not often prove effective and can waste a considerable amount of time and money.

Each project should be bespoke and designed according to the specific question or hypothesis that is being addressed. Once the data requirement has been defined, we then move onto how this is going to be measured. This is often dictated by the Technology Readiness Level (TRL) of the product or level of risk.

If the aim of the trial is the observation of a basic principle or an experimental proof of concept with a high degree of uncertainty regarding outcomes, then it might be more appropriate to engage an external laboratory or academic research facility. These facilities often can complete more detailed sampling procedures and if they have appropriate licenses in place, they are able to trial products or technologies that are currently not legal in our supply chain. An example of this would be the exploration of insect protein meal as a potential alternative to soya. Currently, this is not permitted for use, therefore any evaluation of the effect of feeding insect protein to our broilers would need to be conducted in a research pen trial facility. The approval of insect protein for use in broiler diets is currently being discussed and has significant support. If or indeed when it is approved, we can use the studies already conducted to continue investigating issues, as significant challenges remain to be addressed with the use of insect protein in a relevant environment.

However, depending on the nature of the trial it may not be suitable to conduct work in a pen trial facility. For example, if the aim is to investigate the effect of something that cannot be performed with birds in a shared environment, such as some of our *Campylobacter* or welfare research work, houses on farms are deemed the more appropriate experimental unit. As a product or technology progresses towards market, we then need to assess performance in a relevant environment with more variation or 'noise', e.g. different bird or housing types.

Analysis of data from a commercial setting can be extremely challenging due to the multifactorial nature of the environment; however, advances in mathematics and computing are now able to handle large complex datasets. Often, when we review results, we can be quite fixated on the difference between the means and statistical differences. However, the variance or spread of the data is also crucial and often provides more context. The more variation within a metrics dataset, the more samples you may need to detect significant differences, although within a commercial environment variance can be even more important than the mean as processing facilities need a consistent product to meet customer specifications.

The following chapter summarizes the challenges we as a commercial company have outlined in the context of a case study relating to *Campylobacter* reduction.

7.2 *Campylobacter* and the Dynamics of the Chicken Gut Microbiome

A multidisciplinary consortium between Moy Park (industry), AFBI (public Sector agency) and LSHTM (academia) enabled us to investigate complex research questions in relation to poultry performance and the presence of the enteric bacteria *Campylobacter*.

A constant overarching question evaluated during our discussions was how we can improve poultry performance whilst simultaneously reducing *Campylobacter* numbers, which would have a positive impact along the supply chain on human health. *Campylobacter* is the leading bacterial cause of human gastroenteritis in the world (Gundogdu and Wren, 2020). *Campylobacter* spp. are widely distributed in most warm-blooded animals and the handling and consumption of contaminated poultry is the suggested major cause of infection within humans. When we consider the burden of foodborne diseases globally, where almost 1 in 10 people fall ill each year and 33 million of healthy life years are lost, *Campylobacter* spp. are a major contributor to this number (WHO, 2020). Foodborne disease remains a major challenge for human health (WHO, 2017, 2022).

Campylobacter spp. are found to colonize the chicken caeca with relatively high numbers (up to 10^9 CFU per gram), particularly *Campylobacter jejuni,* which is a leading cause of illness in humans, responsible for 80–90% of cases (Awad *et al.*, 2018). Typically, natural colonization of chickens by *Campylobacter* spp. has been associated with an approximate two-week life cycle. However, we do not understand how, and why, this occurs and particularly what the interface is between *Campylobacter* and the chicken gut microbiome. Currently, we do not understand why *Campylobacter* spp. are typically detected at the approximate two-week point of the life cycle.

Previous studies have investigated how the chicken microbiome develops over time. A rapid increase in microbial diversity in the early days had identified an abundance of *Enterobacteriaceae* that dominate, whilst *Firmicutes* increase up to day seven (Danzeisen *et al.*, 2011; Ballou *et al.*, 2016). We know that the gut microbiome is heavily influenced by the rearing environment and from the diet (Hegde *et al.*, 1982). Studies have investigated how the presence of *Campylobacter* was described to increase *Bifidobacterium* and alter abundances of *Clostridia* and *Mollicutes* (Thibodeau *et al.*, 2015). These studies have improved our understanding of the chicken caecal microbiome in relation to *Campylobacter*; however, what was clear was that if we wanted to try and understand the questions around *Campylobacter* prevalence, a day-to-day study in a commercial setting was required with relevant replicates and metadata. This had not been performed previously. These data would allow us to investigate how the gut microbial population structure evolves over time, and how can this potential impact *Campylobacter* presence. To attempt to answer this question, we performed the first ever day-to-day chicken microbiome study from day 3 to day 35 with 12 replicates per day (final n = 379) with *Campylobacter* detection being performed in a natural environment setting (Ijaz *et al.*, 2018).

Many discussions took place regarding our experimental design. Where should we perform the experiment? Should we infect the chickens with *Campylobacter* or leave to natural infectivity? How would we sample the chickens each day? Were cloacal swabs sufficient for our data analysis? How many replicates would be required to be statistically relevant in terms of microbiome data? After several rounds of discussions, we opted for a natural infectivity experimental set-up, with 12 chicken samples per day, and importantly, we chose to investigate the caeca as opposed to cloacal swab. Our reasoning for this was to perform metagenomics based on the organ that contains the most bacteria within the chicken. The implication of this was that we had to cull 12 different chickens per sampling day and this meant the study design was not a true longitudinal study (as we were not sampling the same chicken over time). The 12 chicken samples per day (essentially replicates) gave us the confidence that we would detect the relative variability between days.

In previous studies, particularly work investigating an intervention, such as probiotic to reduce the presence of *Campylobacter*, the birds were orally 'dosed' with *Campylobacter* (Eeckhaut *et al.*, 2016). Many of these studies report positive impacts of the intervention; however, when tested in a commercial setting the results could not be replicated (Johnson *et al.*, 2017). We believe this is due to the difference between natural and artificial effects of *Campylobacter* colonization and is why we decided to allow natural colonization to reflect the commercial setting more accurately.

Sampling was followed by short-read 16S metagenomics sequencing, which allowed us to perform a variety of statistical analyses including classic microbial ecology alpha and beta diversity metrics to ascertain diversity over time, and how variable the samples were between each other. As expected, we observed a rapidly increasing microbial diversity over the initial days of sampling before plateauing (Fig. 7.1A). To our surprise, when investigating the prevalence of *Campylobacter* within each day, we observed *Campylobacter* in 25% of samples at day 16. However, we did not observe *Campylobacter* again in any other chicken on any other day (aside from one sample at day 3). One possible explanation for this was that chickens that were positive for *Campylobacter* were culled and possibly not allowed to further spread. This was a major discussion and a potential risk with our natural infectivity set-up. We also had the discussion point about why *Campylobacter* appeared again at approximately the two-week time point. Our statistical analyses identified that *Campylobacter* was detected at day 16, which was one day after the most substantial changes in metabolic profiles observed.

Given the lack of *Campylobacter* within our experimental set-up, we used our day-to-day sampling and replicate numbers to investigate how microbial population structure shifts over time and the potential mechanisms for any changes observed. To do this, we had to explore alternative statistical analysis methods, away from the typical alpha and beta diversity techniques used in most microbiome experiments, which, although useful, provide a defined output.

To investigate the factors that drive microbial population structure over time, we characterized the phylogenetic community composition from each sample and analysed whether the microbial community was stochastic (driven by competition among taxa) or deterministic (driven by environmental pressure, i.e. the host environment). We observed a shift from competitive drivers at the early stages of the life cycle to environmental drivers of microbial community structure, specifically from days 12 to 20 creating a window of

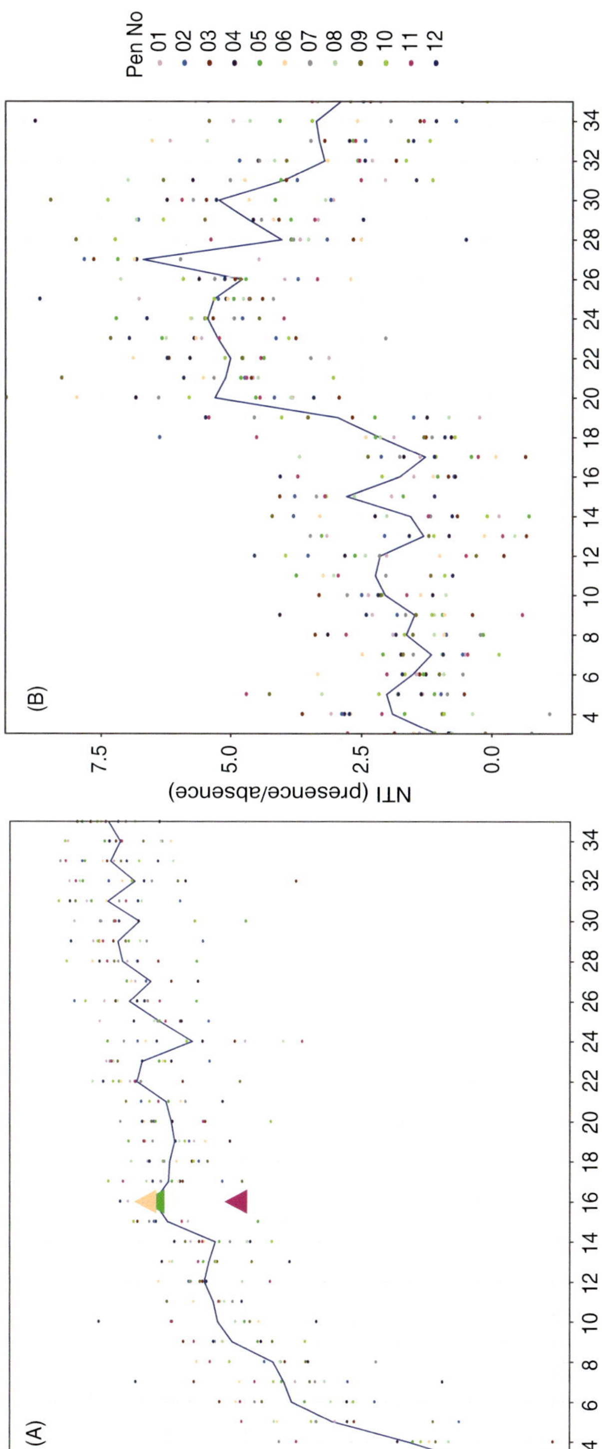

Fig. 7.1. Day-wise statistical measures calculated on the microbiome data. (A) Shannon entropy with first appearance of *Campylobacter* highlighted as triangles. (B) Nearest-taxon index (NTI) and nearest relative index (NRI) considering presence/absence of operational taxonomic units (OTUs) in samples. Samples are coloured with respect to the pens they originate from. (Image from Ijaz *et al.*, 2018). This is an open-access article distributed under the terms of the Creative Commons Attribution License (CC BY).

opportunity whereby *Campylobacter* can potentially appear (Fig. 7.1B). Microbial variation was also significantly influenced by the diet of the chickens, where major differences in taxa were observed upon shifting the diet from starter and grower to finisher.

Overall, even though our study was originally intended to focus on how and why we detect *Campylobacter* within the chicken gut microbiome (particularly at the two-week time point), because of the sampling points and replicates we had accrued, we were able to study how gut microbial population structure evolves over time. One of our findings was the potential window of opportunity where the greatest fluctuation in microbial profiles were observed. These changes are also clearly related to bird age and the changing feed composition. Translating this research into tangible impact may lead to administering feed additives at a much earlier time point in the life cycle as opposed to the current strategy of administration during the latter days. Feed additives are typically added to enhance and/or reduce pathogens such as *Campylobacter* relatively late during the chicken life cycle. Based on our results, there may be logic in administering such additives at an earlier point in the life cycle of the chicken to deter *Campylobacter* infection.

The study also measured metadata in the form of weight and feed conversion rates; however, it was clear we could not analyse the impact of these changes due to an averaging method of our metadata. In addition, based on our statistical analyses, it was clear environmental variables on the chicken gut microbiome were significant and we needed to investigate this more extensively.

7.3 Impact of Production System Parameters on Chicken Microbiomes

Based on our initial study, we chose to investigate three distinct industrial-scale production systems and measure several parameters that could influence chicken gut microbial population structure and performance. In contrast to our previous study, here we sampled fewer chickens but focused more on sampling from different farms that have varying parameters. We investigated 'Normal' (N), 'Higher Welfare' (HW) and 'Omega-3 Higher Welfare' (O) production systems (McKenna *et al.*, 2020). Key differences between the production systems included stocking density and the use of additives such as Omega-3 which have been shown with preliminary studies to enhance growth of chickens. Sampling was less focused on day-to-day variation but more on pre and post-days of when *Campylobacter* typically is detected (specifically day 7 and day 30).

We performed a range of statistical analyses to investigate the impact of measured variables on the chicken gut microbial population structure. As before, initially we performed alpha and beta microbial diversity metrics investigating bacteria that were significantly varied between days. To maximize our ability to interpret the data, we used integrative algorithms such as MINT from the mixOmics package (http://mixomics.org/) to allow classification and discrimination of features across datasets. To explain the variability of microbial population structure between groups, we performed PERMANOVA. However, whilst PERMANOVA analyses show the extent of influence on microbiome population structure in terms of variability, to obtain directions as to whether an increase or decrease in parameters causes an increase or decrease in the properties of microbiome, we performed subset regression on one-dimensional realization of the microbiome (e.g. using our previous alpha and beta diversity analyses).

We identified production system N as distinct from production systems HW and O when comparing the different bacterial genera present within the sampling. In terms of parameters, an increasing stocking density had a decreasing impact in terms influencing gut microbial population structure via environmental pressure. This implies if we place many chickens in a defined space, we have less of an ability to influence chicken gut microbiomes using external strategies (Law *et al.*, 2019). We also identified key genera within the different production systems such as *Eisenbergiella* within HW and O, and likewise *Alistipes* within N, which were representative. Specific genera were identified for individual production systems that have key roles in energy metabolism, amino acid, nucleotide and short-chain fatty acid (SCFA) utilization.

The role of diet is sufficiently important for microbial community structure assemblage as previously described whereby digestion of non-starch polysaccharides (NSPs; found in the grain of chicken feed) lead to production of SCFAs, which are absorbed across the mucosa and catabolized by the host (Jozefiak et al., 2004). SCFAs contribute to chicken nutrition and also lower pH, which can inhibit acid-sensitive pathogens and improve mineral absorption (Sergeant *et al.*, 2014). Thus, there is an association between the consistent and differentiating parameters of the production systems that affect feed utilization, leading to competitive exclusion of genera based on competition for nutrients and other factors. Interestingly, *Campylobacter* was only identified within production system N, which has the highest stocking density. The addition of Omega-3 also increased weight gain; however, we were unable to identify specific taxa that differentiated this from other production systems.

Overall, investigating three distinct industrial scale production systems allowed us to demonstrate the key role of production system parameters in shaping the bacterial communities' impact on the chicken microbiome (Fig. 7.2). Future studies will need to investigate further these extrinsic parameters in conjunction with optimization of diet targeting microbiomes that benefit the chicken in terms of performance and reduce pathogens such as *Campylobacter*.

7.4 Conclusion

Based on our collaborative research efforts, we have been able to contribute significantly to the knowledge of the broiler microbiome by observing the evolution of the microbial species diversity over the lifetime of the bird and in different management systems. However, the dilemma still persists that those changes in microbiome stimulated by, for example, probiotics, do not consistently result in differences in commercial key performance indicators e.g. feed conversion ratios, liveability or body weight gain.

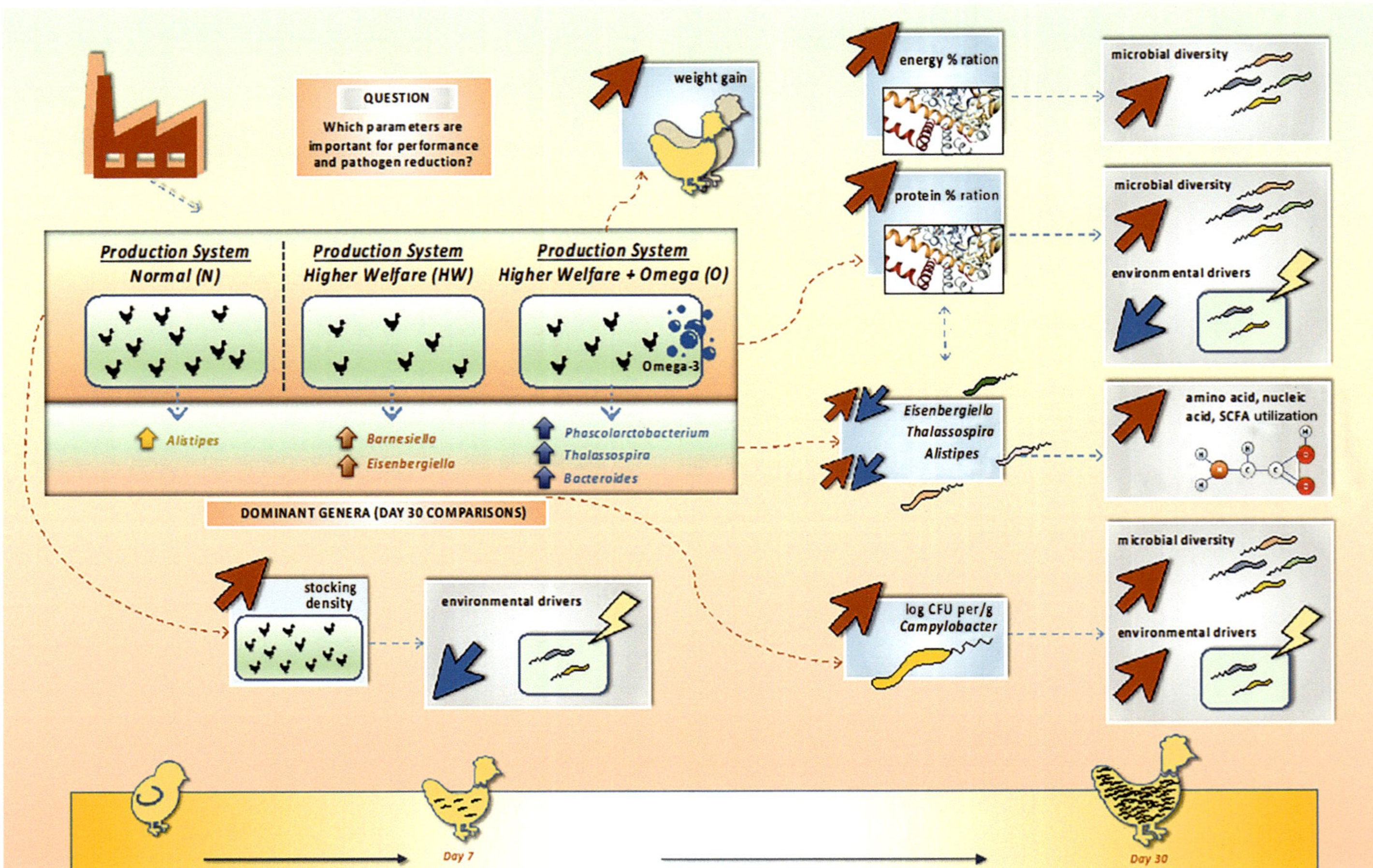

Fig. 7.2. Impact of industrial production system parameters on chicken microbiomes (McKenna *et al*., 2020).

The key to unlocking the power of the microbiome to positively influence the health of broilers remains elusive, although we remain positive that new 'omics' techniques will help further our understanding.

References

Awad, W.A., Hess, C. and Hess, M. (2018) Re-thinking the chicken-*Campylobacter jejuni* interaction: a review. *Avian Pathology* 47(4), 352–363.

Ballou, A.L., Ali, R.A., Mendoza, M.A., Ellis, J.C., Hassan, H.M. *et al.* (2016) Development of the Chick Microbiome: How Early Exposure Influences Future Microbial Diversity. *Frontiers in Veterinary Science* 3, 2.

Danzeisen, J.L., Kim, H.B., Isaacson, R.E, Tu, Z.J. and Johnson, T.J. (2011) Modulations of the chicken cecal microbiome and metagenome in response to anticoccidial and growth promoter treatment. *PLoS One* 6(11), e27949.

Eeckhaut, V., Wang, J., Van Parys, A., Haesebrouck, F., Joossens, M. *et al.* (2016) The probiotic *Butyricicoccus pullicaecorum* reduces feed conversion and protects from potentially harmful intestinal microorganisms and necrotic enteritis in broilers. Frontiers in Microbiology, 7, 1416. doi: 10.3389/fmicb.2016.01416

Gundogdu, O. and Wren, B.W. (2020) Microbe profile: *Campylobacter jejuni*–survival instincts. *Microbiology*, 166(3), 230–232.

Hegde, S., Rolls, B. and Coates, M.E. (1982) The effects of the gut microflora and dietary fibre on energy utilization by the chick. *British Journal of Nutrition* 48(1), 73–80.

Ijaz, U.Z., Sivaloganathan, L., McKenna, A., Richmond, A., Kelly C. *et al.* (2018) Comprehensive longitudinal microbiome analysis of the chicken cecum reveals a shift from competitive to environmental drivers and a window of opportunity for *Campylobacter*. *Frontiers in Microbiology* 9, 2452.

Johnson, T.J., Shank, J.M. and Johnson, J.G. (2017) Current and potential treatments for reducing *Campylobacter* colonization in animal hosts and disease in humans. *Frontiers in Microbiology* 8, 487. doi: 10.3389/fmicb.2017.00487

Jozefiak, D., Rutkowski, A. and Martin, S. (2004) Carbohydrate fermentation in the avian ceca: a review. *Animal Feed Science and Technology* 113(1–4), 1–15.

Law, F.L., Zulkifli, I., Soleimani, A.F., Liang, J.B. and Awad, E.A. (2019) Effects of reduced-protein diets supplemented with protease in broiler chickens under high stocking density. *Animal Production Science* 59, 2212–2221. doi: 10.1071/AN18477.

McKenna, A., Ijaz, U.Z., Kelly, C., Linton, M., Sloan, W.T. *et al.* (2020) Impact of industrial production system parameters on chicken microbiomes: mechanisms to improve performance and reduce *Campylobacter*. *Microbiome* 8(1), 128.

Sergeant, M.J., Constantinidou, C., Cogan, T.A., Bedford, M.R., Penn, C.W. *et al.* (2014) Extensive microbial and functional diversity within the chicken cecal microbiome. *PLoS One* 9(3), e91941.

Thibodeau, A., Fravalo, P., Yergeau, E., Arsenault, J., Lahaye, L. *et al.* (2015) Chicken caecal microbiome modifications induced by *Campylobacter jejuni* colonization and by a non-antibiotic feed additive. *PLoS One* 10(7), e0131978.

WHO (2017) *Diarrhoeal disease*. Available at: https://www.who.int/news-room/fact-sheets/detail/diarrhoeal-disease (accessed 30 March 2023).

WHO (2020) *Campylobacter*. Available at: https://www.who.int/news-room/fact-sheets/detail/campylobacter (accessed 30 March 2023).

WHO (2022) *Foodborne diseases*. Available at: https://www.who.int/NEWS-ROOM/FACT-SHEETS/DETAIL/FOOD-SAFETY (accessed 30 March 2023).

CHAPTER 8
Reviewing the Literature on Feed Additives: Considerations in Data Collection and Data Analysis

Tom Dale* **and Mike R. Bedford**

AB Vista, Marlborough, UK

Abstract

Providing it is implemented correctly, any single scientific study provides information that hopefully supports or refutes a hypothesis being tested. Multiple studies in a subject area often deliver a range of responses and hence consensus is often hard to achieve. The equivocal nature of the responses noted between similar studies often relates to the variance in conditions under which each study was conducted. Any trial result is subject to influence from several factors, some known and some not. While the known factors should be reported (and often are not), the unknown factors may or may not be reported and as a result the scientist is presented with what looks like a disparate data set. The scale and complexity of the data available in many subject areas can be overwhelming. Most reviews are the authors' subjective interpretations of the papers that they have selected from the total available, which is almost always only a very small proportion of what is available. For example, a Google Scholar search on the topic 'Phytase AND Broilers' resulted in more than 23,000 hits, which highlights the scale of the problem when interpretation of the whole literature is the goal. Data-driven reviews encompassing all information in the literature enable hidden conditions to be revealed. Meta-analysis encompasses diverse methods of data collection, selection, analysis and outputs, ranging from simple averages to more complex statistical methods. The advantage of meta-analysis is that careful pre-selection favours more homogenous data sets with focused and narrow models generated as a result. Holo-analysis was introduced and defined as 'the integration of all available data on a specific subject quantifying a dependent response in terms of all available genetic, chronological,

*Corresponding author: Tom.Dale@Abvista.com

DOI: 10.1079/9781800622746.0008

environmental, geographical, managemental, dietary ingredient and nutrient independent variables'. As opposed to meta-, the prefix holo- is clearly definable, entire. Data-driven reviews may well find factors that influence the response that were never recognized in any of the individual trials conducted. For example, the beneficial effect of the inclusion of a coccidiostat in poultry diets on the scale of response to a dietary phytase was revealed using a holo-analysis of all the relevant phytase literature even though no individual trial addressed this question. As more of the 'unknown' factors are revealed, it should be incumbent upon scientists working in the subject area to provide these key data in any subsequent research so the value of the literature can be optimized.

8.1 Introduction

In the animal-feed industry, and that of all scientific fields, evidence is a requirement for successful advancements in technology to produce the best possible products for the market. This is especially relevant today in a world of disinformation rife with fake news. Science must always be underpinned by transparency and the scientific method is currently the most effective tool available to seek truths and solutions to problems that the animal-feed industry faces.

The scientific method allows us to formally assess the significance of a relationship between different variables or groups. Before experiments begin, researchers make predictions using hypotheses. The null hypothesis always predicts no effect, no relationship between variables or no difference between groups. The alternative hypothesis is the researchers' predictions of an effect occurring by a relationship between variables and/or a difference between groups. The assumption is that the null hypothesis is always true and statistical analysis is employed to determine the probability of obtaining results under this assumption. The null hypothesis can then be accepted or rejected.

The researchers then attempt to discuss their results in the context of the entire body of evidence, citing other scientific works and how they support or refute their claims. Therein lies a problem, given the density and complexity of feed additives literature, where a Google Scholar search on the topic 'Phytase AND Broilers' results in more than 23,000 hits. No individual scientist or group of scientists could ever give an objective interpretation of the entirety of the literature surrounding a specific topic. Indeed, most scientific paper discussions and literature reviews are the author's or authors' subjective interpretations of the small subset of papers that were selected. In defence of this fact, research findings are peer-reviewed by others, who determine using their subjective knowledge and opinion if the results were obtained by appropriate methodology, the reporting was transparent and that the claims made by the researchers were supported by their data.

For these reasons, a component of all scientific papers and reviews will be subjective. This necessitates the identification of an objective methodology for the robust analysis of all the available data.

8.2 The Need for Analyses

There is one question underpinning all research in the feed additive literature: when does feed additive 'X' give the most effective response? To answer this question, the interactions between different feed additives, raw materials and feed technology must be understood.

In 2001, Marquardt and Bedford regarded enzyme use as thus far infantile, short of full potential, notwithstanding the then current highly beneficial applications. In 2022, one could draw similar conclusions. As Rosen (2006) summarized, the problems to be solved included:

1. How to select an enzyme from the large multitude on offer.
2. The need for meaningful assays or measurements to determine not only the amount present in a feed but also its biological relevance.
3. Relative evaluation of the expression of enzymes in plant raw materials compared with their exogenous application.
4. A need for better predictions than those from simple linear models.

Four areas for future research and development included:

1. More biologically meaningful assays.
2. Fitting enzyme properties to animal species, substrates and environments.
3. Better understanding of animal responses in cereal-based diets.
4. The use of models to predict enzyme responses and thereby optimize economy and efficacy of usage.

Whilst there has been progress on the points above, there are still many unknowns and, to answer any of these questions, one cannot rely on any individual scientific paper or even draw from a large collection of papers saved in an individuals' reference management software to solve them.

More than 20 years ago, Partridge and Bedford (2001) stated: 'The challenge is to find methods of predicting enzyme response so that enzyme application in all classes of livestock becomes increasingly a science rather than an art'. This was agreed by Choct (2001): 'The concept of predicting the effect of enzymes in a particular feed is attractive, because the producer could then adjust the enzyme amount and/or the nutrient specifications in diet formulations'. This underlines the need for data-driven reviews that allow 'real world' predictions of a feed-additive efficacy.

8.3 The Difference Between Meta-analysis and Holo-analysis

8.3.1 Meta-analysis

By far the most common method of pooling multiple scientific studies for the analysis of multiple analyses is meta-analysis. Coined by education researcher Glass (1976), he described meta-analysis as 'a rigorous alternative to the casual narrative discussions of research studies', aiming for 'the analysis of analyses and the statistical analysis of a large collection of results from individual studies for the purpose of integrating the findings'.

There is a certain level of ambiguity to that statement. By using the descriptor 'large', one cannot attempt to accurately quantify just how much data should be included or excluded from this form of analysis. By definition, meta- is also ambiguous, simply defined as '*higher; beyond*' in the Oxford dictionary and meta-analysis as '*research that combines the results of a number of related studies*'. Meta-analysis then, without absolutes, may be more of an art than a science with, for example, only two trials analysed together having the potential to be labelled a meta-analysis. As previously mentioned, this is the exact problem that Bedford and Partridge identified in 2001.

There are clear advantages to meta-analysis for publication purposes. Researchers make several decisions regarding the study search criterion applied – the objective criteria for study inclusion or exclusion, how they manage incomplete data and which analysis to use on the data. This careful and methodical pre-selection often has a beneficial effect on creating homogenous data sets that generate narrow, focused models. This can have detrimental effects because researchers can, in principle, apply multiple rules for all the above-mentioned tasks, find the perfect study inclusion rules to cherry-pick the data sets they desire and therefore, find the analysis that is the best fit for their hypothesis, resulting in bias. Such pre-selection often results in a more homogenous data set and, as such, the models generated are often more focused and narrower, with excellent R-square (R^2) and root-mean-square error (RMSE) values to describe the proportion of variation in their model.

Often, such a stringent criterion results in only a minority of the total data being considered in any research area. This creates an inherent limitation to the generated models; their ability to predict responses under a variety of conditions is limited by the variation in the data set considered.

8.3.2 Holo-analysis

In contrast to meta-analysis, holo-analysis is a self-explanatory descriptor. It is a holistic analysis, giving more certainty and precision in its description than the progenitor, meta-analysis. First described by Rosen (2004), holo-analysis is 'the integration of all available data on a specific subject quantifying a dependent response in terms of all available genetic, chronological, environmental, geographical, managemental, dietary ingredient and nutrient independent variables'.

The concept emerged due to legislation changes in the European Union (EU) in 1999, resulting in a ban on the use of antibiotics in feed without veterinary prescription. This triggered an urgent demand for alternatives to antibiotics within the feed industry. This was a foreshadowing of the magnitude of the task required to investigate the efficacy of alternative products, such as enzymes, microorganisms, acids, botanicals, yeasts and derivatives, oligosaccharides, aromatics, etc. Prior to this rule change, the evaluation of antibiotics was largely performed by meta-analysis and as Rosen described:

> Not all available independent variables were evaluated, with main attention restricted to the influence of dosage, level of negative control performance, duration, year of test, the influence of anticoccidials and the impact of diagnosed diseases… The concept of holo-analysis emerged from a long series of evaluations of the efficacy of 15 different antibiotics in animal nutrition, encompassing the evaluation of a total of 12,153 negatively controlled tests from 1949 to 1991 in 4301 reports, using data from 55 countries. (Rosen, 1995)

Holo-analysis allows the translation of results in research conditions into field conditions, e.g. optimal vs sub-optimal dosages, pellet vs mash feed, high vs low phosphorous, high vs low protein, etc. With holo-analysis datasets, responses can be predicted with confidence limits and software can be utilized to quantify responses that are unique in time, place and economics.

The identification of missing variables through holo-analysis is one of its greatest strengths. Holo-analysis models can predict effects that have not been directly tested in any individual papers used, which increases the value of already published data and requires no extra animal experimentation. For example, most research trials will monitor feed conversion ratio (FCR), yet do not monitor the effects of individual components of the diet on their results, e.g. wheat inclusion level, pen temperature and humidity, total phosphorous level, etc. The discovery of unpredictable independent variables in a wide range of conditions is a powerful tool, found only with holo-analysis.

Owing to the all-encompassing inclusion of data with holo-analysis, the models generated are often less precise, but usually more robust in nature, making them applicable in the entire spectrum of conditions that exists within the literature used. This is a defining characteristic of holo-analysis; it will be more applicable than other methods in real world situations. Models generated from holo-analysis can predict the optimum dosages of exogenous enzyme products and compare the efficacy of different products to establish the best competitive product (providing sufficient data is available) whilst providing confidence limits for the estimate thereof. It is necessary to make considerations when collecting, selecting, utilizing and analysing data in holo-analytical models. The following section covers some key aspects of holo-analysis, much of which has been previously described by Rosen (1995, 2001a, b).

8.4 Data Collection

All data can be collected from the large pool available within the scientific literature. The models generated from these data should then provide insight into the entirety of the field of interest from which the data was mined. This underpins the importance of the initial data collection stage, where all subsequent work is supported from this base of data. Without appropriate data collection, it will not be possible to accurately quantify the relative contributions and interactions of treatment, feed, genetic, environmental and management independent variable factors on the response variable of interest to enable a prediction of such responses on manipulation of the significant variables identified. There are several considerations to make during the data collection process to ensure the most robust and reliable models are generated.

First, a broad criterion for keywords must be chosen in the search strategy that accounts for the often poor and inaccurate titles, keywords and abstracts seen in publications. A poor selection of keywords will result in data being involuntarily excluded from the holo-analysis. If researchers can bring clarity and more uniformity to title and keyword selection, the data they generate will be able to be used to their full extent in the future.

It is critical to identify and eliminate any data that has been duplicated (for example, when three different publications have examined the same data). This is necessary to prevent unconscious bias or the weighting of any one data set. It can be particularly tricky to identify duplicated data that have small divergences in the data between duplicate publications but identifying and removing them is essential for a representative analysis.

During data collection, the variables that should be included must be identified. The best approach, although most time consuming, is to collect all possible variables from every paper to create the most holistic data set. This also reduces decision making in choosing which variables to add and/or exclude, which reduces bias in the data set. This does not mean that all variables will be included in the analysis; however, it gives the most options for subsequent models generated. By recording all possible variations, it is possible to identify surprising associations with the production variable of interest that would otherwise go undiscovered. It would be prudent to scrutinize a sample number of publications that will be used in the analysis to identify all the possible variables that can be extracted from the papers to ensure none is overlooked.

When selecting data, it is important to always reflect if it is biased and therefore not meaningful. As previously discussed (Bedford and O'Neill, 2016), the 'meaningful' nature of the data relates to the uniformity in the detail of the materials and methods employed in each study. Despite suggested minimum guidelines for scientific papers, e.g. the Poultry Science Association guidelines or more general guidelines, such as the 'Gold Standard Publication Checklist' (Hooijmans *et al.*, 2010) or the ARRIVE guidelines for biomedical research (Kilkenny *et al.*, 2010, 2014), there is mixed adherence to this advice in the literature. It is often apparent that the minimum guidelines set out by a journal do not possess sufficient detail for the trial to be repeated or even included in a systematic review (Hooijmans *et al.*, 2010). For illustration, in a survey of 271 papers on '*biomedical research on laboratory animals*', only 13% of the papers simultaneously reported the weight and the age of the animals used (Kilkenny *et al.*, 2009).

This is a particular problem in speciality topics such as feed enzymes (Rosen, 2006), where more specific guidelines are needed and available, yet rarely referred to. This results in gaps in the details available for some of the key variables that influence responses to a given factor. Whilst attempts have been made to contact the authors responsible for those papers (Rosen, 2000), this has been found to be time-consuming and not guaranteed to improve the data set quality.

The problem is apparent in the feed enzyme dataset where significant numbers of publications do not adhere to the standardized naming of products or their units of activity. This often make it impossible to correctly identify and classify some products. Rosen (2000) stated the enzyme resource yielded 252 different generic descriptors of enzyme activity, with very few being recognized methods of nomenclature and today reporting is still insufficient. It is not uncommon to see enzymes named generically as 'carbohydrase' or 'pentosanase', which do not meaningfully define the enzyme in question. More biologically relevant, simple nomenclature is necessary to generate the most accurate and significant models by reducing the number of variables within the variable 'enzyme'.

A bare minimum standard should be established in all journals when reporting feed enzymes for publication, which would include the source organism and production strain to understand better the gene from which the enzyme was transcribed, along with the organism that hosted the gene and produced the enzyme. This information, particularly the source organism, influences the characteristics and efficacy of the enzyme. Even with this minimum information, the source organism can produce several isozymes of the described activity that may also differ significantly so more information would be optimal.

In many cases, the product in question is not pure and will contain multiple side activities from the production strain, which may or may not be related to the enzyme of interest. These side activities can be just as crucial as the main declared activity (if even declared). Even when the presence of main activities and side activities is declared, an entirely new problem is that the units of activity are often not declared properly. First, the activity level may not be declared at all and second, the units given can often prove another point of frustration if the use of alternate or unpublished methods in some publications is used despite perfectly serviceable and internationally accepted methods being available. This often renders the data worthless and the relevance of the enzyme to the outcome unknown. One potential solution here would be an attempt at providing a clear conversion factor between the established and proprietary assays. When this is not achieved, the ability of the subsequent analysis to extract meaningful insights, particularly with regards to determining optimum dosage and product comparisons, is severely constrained.

Whilst this example is specific to enzymes, the underlying principle here is that any scientific study should be presented in a way that allows replication by others solely from reading the paper alone. When shortcuts are taken in reporting, any output is deficient because of the inaccuracies and inconsistencies, which means the trial could never be truthfully repeated. This impacts the utility of post hoc analysis in all fields, where the product of interest suffers from poor adherence to nomenclature. This includes all feed additives, where huge benefits could be found from more consistent reporting and stricter editorial demands regarding compliance.

8.5 Data Entry, Storage and Organization

Critical to a successful holo-analysis, yet often overlooked, is data entry. Due to the variable nature of scientific papers, much of the data will need to be manually mined from each individual paper by the user. This requires immense attention to detail to avoid typical data entry errors, as there is a significant amount of human input required to enter the data. Common errors include transcription errors, such as typographical errors, repetition, deletion and other human errors. These include spelling mistakes (see data cleaning section), accidentally hitting an extra key and adding an unwanted digit or missing out a digit entirely, all of which disrupt the accuracy and validity of statistical analysis. Transposition errors are also common when manually inputting data to spreadsheets, where the position of characters is unintentionally switched, for example, 0.193 may be incorrectly inputted as 0.913, which can have severe consequences for a data set. To minimize errors of this kind, using the copy and paste function for as much data as possible is advised; however, this can set off a new kind of data entry error, incorrect data formatting. It is not uncommon for feed enzyme holo-analysis spreadsheets to contain over 250 columns and 2000 rows of data. It is easy to paste data into the wrong field entirely and thus create erroneous data that can throw off an entire analysis. Care must be taken when inputting data, as no matter how robust and strong a statistical model is that is generated from a data set, if the data used were incorrect, the entire model will be invalid. Some errors will be picked up during the data cleaning procedure (see below); however, some smaller errors will not be picked up, which is why so much care must be taken. For example, in a phytase holo-analysis data set, one may simultaneously record overall phosphate levels, total phosphorous (tP), actual analysed phosphorous (analysed P), non-phytate phosphorous (npP) and digestible phosphorous (dgP). It is not difficult to accidentally add the npP value (e.g. 0.55) in the dgP (e.g. 0.48) field and vice versa, which will not be picked up by distribution analysis and seem to fit the data trends perfectly, yet are still incorrect and add to the inaccuracies in the analysis and the eventual model generated.

Although more complex methods may be available, a spreadsheet or simple relational database is more than a sufficient means of data storage. More complex methods may be cumbersome and lead to greater difficulty in analysing the final data set. The most straightforward way of getting data ready for analysis is to enter it in a linear manner, with each spreadsheet row representing all the information relating to the outcome of one treatment.

The output will be an independent variable interest which is an absolute performance metric such as body weight gain, FCR or European Production Efficiency Factor (EPEF). Rosen (2000) revealed that to achieve the most robust models for real-world analysis, the data should be entered so that each row represents the effect of the treatment compared with the control, rather than just the absolute performance of all treatments, including the control. In a spreadsheet this could look like the following formula (FCR Control value 'F2' – treatment value 'F3' = FCR Effect 'G3'). Using this value rather than the absolute values draws the control performance into any model as a significant variable, as it is a measure of the performance of the animals in the trial of interest for their given age or weight, resulting in more robust models, notwithstanding a lower R^2 and larger RMSE. Care must be taken to only include trials with a true negative control for the product in question to generate the 'effect' value of control minus treatment for the most robust models. See Table 8.1 for an example format.

8.6 Data Cleaning

After the data has been selected and entered into a suitable database, its accuracy must be established. The simplest way to do this is to plot out the distributions of each variable, which is an informative process, available in most statistical packages. This will generate means and standard deviations for each variable, which give useful and descriptive statistics along with distribution graphs that give an initial suggestion whether or not the data are normally distributed and any clear anomalies should be obvious. Still, data are sparse and therefore valuable in animal production data sets and should only be excluded if it is unambiguously clear that the value is erroneous. It is tempting to exclude any data significantly different from the mean, yet, in doing so, the variation in the data set is lost and can lead to the subsequent model failing to detect some associations of interest. On the contrary, it is possible that very few erroneous data points are required to significantly reduce the likelihood of obtaining a robust model. This is where you may discover some of the larger transcription errors, where a value may be ten times or a hundred times larger or smaller than the rest of the values within the variable data, if a decimal point has been misplaced or a digit added or subtracted from the true number. In these instances, the values should be re-examined from the original paper and in some instances the values may not actually be erroneous at all. Therefore, care should be taken in this process because, as previously inferred, the quality of the model is dependent upon the quality of the data. An example can be seen in Table 8.2, where ten intentional errors have been added to a dataset of 256 observations looking at the effect of phytase supplementation on the bone ash percentage of broilers. These errors account for just 3.8% of the data in a single column for net bone ash effect. The negative effects on R^2 and RMSE can be clearly seen in the model generated from the erroneous data.

Attention must be given to ensure that all values are correctly aligned with one another. For example, the misidentification of a 21-day body weight as that of a 42-day body weight in the column for body weights would introduce errors into the 42-day analysis but not necessarily be caught in a simple distributions test as mixed age data are present in the weight column but not age-sorted.

With class variables, both discrete and nominal, there is the potential for more transposition errors to be identified, by running the above-mentioned distribution analyses, it is possible to sort them alphabetically and enable the discovery of incorrectly entered character errors. A single character misplacement will produce two discrete and distinct variables from the intended singular variable, which can have negative ramifications on the analysis, especially if this occurs more than once. Examples include names of feed additives where 'xylanase and 'xylanasse' (a misspelled name) would be considered as different products. Any column with characters and not digits is susceptible to this kind of error and care should be taken to identify these during the data cleaning process.

Table 8.1. Example showing the suggested formatting for holo-analysis data organization.

Manuscript	Country	Strain	Treatment	Age at start of experiment (days)	Age at end of experiment (days)	Control 42-day FCR	Cumulative 42-day FCR	FCR Effect (treatment – control)	% change in FCR
Johnson *et al.*, 2019	UK	Ross 508	1	1	42	1.59	1.59	0.0	0.0
Johnson *et al.*, 2019	UK	Ross 508	2	1	42	1.59	1.57	–0.02	–1.3
Johnson *et al.*, 2019	UK	Ross 508	3	1	42	1.59	1.63	0.04	2.5
Johnson *et al.*, 2019	UK	Ross 508	4	1	42	1.59	1.42	–0.17	–10.7
Anderson *et al.*, 2021	USA	Cobb 500	1	1	42	1.67	1.67	0.0	0.0
Anderson *et al.*, 2021	USA	Cobb 500	2	1	42	1.67	1.53	–0.14	–8.4
Anderson *et al.*, 2021	USA	Cobb 500	3	1	42	1.67	1.73	0.06	3.59

Table 8.2. Example of the negative effects of erroneous data on the R2 and RMSE of models generated with holo-analysis.

Error number	Original data value	Erroneous data value
1	6.96	–69.6
2	10.04	–100.04
3	16.28	–16.28
4	11.48	–11.48
5	15.68	156.8
6	12.74	122.74
7	5.4	54.0
8	8.9	89.0
9	7.2	72.0
10	9.9	99.0
R^2	0.495	0.057
RMSE	2.423	17.282
Prob > F	<.0001	0.0860

8.7 Selection of Variables for Use in Models

Deciding which variables to consider as independent or dependent is straightforward when there are limited variables available. It is usually known that specific independents correlate with specific dependents, meaning the only challenge is to ensure that the independents are not correlated with one another during model fitting.

Sometimes, variables may appear independent when they are not. One such example is when the dependent variable is calculated from independent variables. The European Production Efficiency Factor (EPEF) value is not independent from the values for FCR, body weight, etc. that are used to calculate it. However, both variables are most likely dependents in this instance and not likely to be included as model variables, although the principle still applies. This can be seen in survey-based data and caution should be applied when dealing with it.

In holo-analysis, there are likely to be large numbers of variables. The task here is to identify those that are most important. Prioritization should be given to generating models with as few variables as possible. The first and obvious reason for this is to prevent overfitting of the model. The second is practical and relates to interpretation. The more variables and the more interactions included in a model will result in a model that fits the data well but makes practical understanding or application complicated and unintuitive.

During the initial screening of possible variables, the selection process should labour to only select variables that are not correlated with one another. Utilizing a principal components analysis is a useful method of screening and assessing correlations between variables, concurrently. Variables lying within the same principal component cluster should be considered correlated and hence the most highly correlated with the principal component identified as a potential variable of interest.

When the dependent variable is in a principal component and has a high correlation component with a group of independent variables, the most highly correlated independent variable should be considered for inclusion in the model and the remaining variables excluded.

Through our experience working with commercial data sets, the number of principal components should ideally number ten or less, to limit the size of the model as mentioned above. Caution is urged if an independent variable does indeed explain some of the variation in the independent variable as it could be behaving as a proxy for the excluded but correlated independent variables. One such example is that many dietary amino acids correlate with one another and, resultingly, only one can be selected. This can create an issue between statistics and real-world knowledge. Perhaps methionine is the highest-correlated variable statistically, yet it is recognized that lysine is most relevant from a biological perspective. Given these circumstances, the statistically favoured independent variable can be replaced by the more biologically and commercially relevant variable. For statistical reasons, there are several cases where the biological or commercial relevance of a selected variable may need to take precedence over the variable selected.

The above processes are helpful at selecting independents linearly related with the dependent. In many instances it is prudent to look at polynomial or non-linear relationships by plotting all potential independents against the dependents to visually determine if there is an alternative relationship to be found. This is a time-consuming process but can provide great value. After identification on this basis, they should be subsequently tested for correlation with the methods described above before a model is constructed.

Once visual relationships are established, the next challenge is to identify the appropriate term (e.g. quadratic, non-linear, logarithmic) for the selected variable actually predicts the dependent in the subsequent models.

8.8 What Makes a Good Model?

The necessities of a good model are that it describes a large proportion of the variation in the data set whilst having high degree of confidence in any predicted value determined. This is achieved with a high R^2 and low error term, respectively.

This is an oversimplification as a good model should be populated with variables that are truly causative, which is difficult and sometimes impossible to prove. A golden rule of statistics is that correlation does not imply causation and, consequently, models must be interpreted accordingly. For example, there is a highly significant relationship between margarine sales and divorce rates in Maine, USA, with an R^2 value of 0.98 (Fletcher, 2014). A mere correlation, however, does not mean that one causes the other but purely coincidence and should be a warning for the interpretation of any models generated.

Another important realization is that regardless the magnitude of any data set achieved, it is probably only a fraction of the total available data. This should be a foundation in understanding that any model produced is merely an estimation of the model that could be generated should one have all available data included. This is what is known as model robustness. A term that describes the ability of the model to apply to data that it has not been exposed to. In this hypothetical situation, the definition of a robust model would be that it is able to accurately predict the absolute values of a data set where (n = all).

Practically, robustness of a model is tested by its utility over time. If a model generated in 2017 can be applied to the data of 2022 and still provide a relatively good fit, the model can be considered to have a degree of robustness. In principle, if a model were truly robust, the variables involved would not change over time and nor would their coefficients once more and more data become available.

A tool is available to test robustness of models. It involves the segregation of the data set into separate pools. One portion is used to generate the model and one or more random pools are utilized to test whether the model generated is a good predictor of the held-back data sets. The model generated from the first pool of data is applied to the remaining naive pools of data and tested to see if it can predict the independent with a reasonable degree of precision. When the R^2 values and error terms that are generated by the model on all data sets are very similar, it can be ascertained that the model is robust. Robustness should be the goal in any model, not merely a high R^2 and low error rate on one individual dataset.

Whilst the number of data points necessary to achieve a robust model depends on the data set at hand, it can be generally assumed that a larger data set is more likely to reap success as many data sets are not large enough to split into data sets for testing and validation.

In principle, the fewer errors a data set possesses, the more the data set is aligned with reality and therefore the smaller the data set can be while still generating a robust model so long as the variables present in the data set have a true role in the prediction of the independent.

On the latter point, it is beneficial to understand the physiology and potential mechanisms at play in driving the response of interest. This allows for biological knowledge and understanding to help select potential variables in the model but this is not always easy to implement. An example of this is the notion that air quality may impact broiler flock performance and that it may be desirable to include this variable in performance models (Beker *et al.*, 2004; Donaldson *et al.*, 1995; Reece and Lott, 1980). However, in our experience collecting data commercially, it is not possible to acquire such information consistently from every farm.

Further, if there are no independent variables present that are relevant for the estimation of the independent then there is no data set, no matter how large, that will generate a meaningful model. Rosen (2003) noted that relatively few lines of data are necessary in some situations:

- 20 data points were required for reasonable first efficacy appraisal of an additive.
- 50 data points were the minimum to establish the key variables influencing the response.
- More than 300 data points were needed to provide reliable models from which commercial decisions could be made.

There is clearly a minimum number of rows of data that are needed to generate a 'reasonable model' and a robust model will often require hundreds if not thousands of data points. The eventual goal should be the creation of a robust model, a model that no longer evolves with the addition of further data.

8.9 Model Types

There is no single best model when investigating the relationships between dependent and independent variables. However, the identification of the optimum coefficients for each of the independents that are most influential on the outcome of interest and are key in determining the optimum output should always be the goal.

8.9.1 Linear and quadratic models

Models are frequently based on linear terms and sometimes extends to quadratic terms. The models also usually have interactions between some, if not all, of the terms included in the model (Hooge *et al.*, 2010; Letourneau-Montminy *et al.*, 2010; Rosen, 2002; Sales, 2014).

When dependents are merely linearly related to the independent, the model continues to improve with every increment or decrement in the independent. There is no predicted maximum or minimum with linear models. The maximum output is achieved within the confines of the data set and it will be associated with a very large confidence limit because of the limitations in data at these extremes. A further consideration is that biological systems are not usually linear in response to an independent variable when the range used is wide enough, so the extension of the model beyond the data set is not advised.

Quadratic terms can be implemented for independent variables when the response to linear increments of the variable asymptote and then decline when a maximum has been achieved (Letourneau-Montminy *et al.*, 2010; Siegert *et al.*, 2015). This is a common biological effect of nutrients that can become toxic when fed in excess. When quadratic terms are assigned, the rate that performance increases towards its optimum with increasing dose of an independent mirrors also the reduction in performance with incremental dosages beyond the optimum. This is because the relationship assumes symmetry in the approach to both the maximum and minimum. Thus, careful consideration is required and a good biological justification necessary when employing quadratic terms, given that, in many cases, nutrients improve performance up to the optimum but without reducing through further increments beyond the optimum dosage. For example, the animal may excrete or re-route the nutrient in excess to elsewhere in the body without consequence on the dependent variable. In some cases, the excretion or detoxification process will reach a point where it burdens the dependent variable of interest and a decline in performance may be observed.

Broken-line models or broken-line quadratic models are alternatives to consider when addressing the above issue, which assume no further movement in the dependent variable once asymptote is achieved (within the constraints of the independents).

8.9.2 Non-linear models

There are a range of non-linear models and, through understanding their biological relevance, specific non-linear models can be applied (Sauer *et al.*, 2008). Two options are asymptotic and exponential models; they can be considered against the broken-line models previously mentioned, with caution that these models reach optimum response at an infinite dose of the independent variable. This means calculation of the optimum dose is irrelevant along with the biological relevance of the model generated as the independent variable's dose increases beyond 'normal' ranges. As a result, such models need to be interpreted carefully. Regardless, a model still must be statistically significant to be considered relevant, which means that a model with biological relevance that does not fit the data set well should be discarded and alternatives should be considered.

8.9.3 Partition models

Partition models, such as random forest and boosted tree, allow simple methods of combining aspects of different modelling types all in one. There are both simple and sophisticated versions of partition models and both should be considered.

Depending on how the data set is sequentially split (to improve the statistical fit), the shape of the relationship can be linear, quadratic, asymptotic or threshold. Based on how many splits are introduced to the model during partitioning, there will be a varying number of 'lines' created and this is a component that is difficult to predict prior to analysis. Given their coarse nature, the output from partition models should only be used as a tool in determining the variables, functions and interactive terms that may be considered in linear or non-linear models.

8.9.4 Neural network models

In animal performance data sets and in the prediction of nutrient contents of ingredients, neural networks are a method often producing the highest R^2 value and lowest errors terms in their models (Mehri, 2013; Perai *et al.*, 2010; Savegnago *et al.*, 2011).

Unfortunately, the number of nodes and relationships between them makes it unclear whether the specific setting selected has given the true optimum possible model. This problem is exacerbated by the fact that even with fixed settings, re-running the same model will arrive at a different solution since each run presents a unique solution. One further issue with neural network models is that, when compared with all other forms of modelling that are easily transcribed into other media for display of the output, there is inherent difficulty in parametrizing the output in a manner that the solution can be used to explore the interrelationships between the independent and dependent variables in a spreadsheet. Given that this is often the desired endpoint during analyses, it results in neural network models often falling short.

8.10 Modelling Considerations

The power of holo-analysis is best realized when investigating the effect or value of a specific nutrient or additive in the diet. It has been previously elaborated that holo-analysis, considering all available data for inclusion in analyses, often results in poorer R^2 values and error terms compared with those achieved with meta-analysis driven models due to the carefully selected data and variables of the latter.

Most of the variation in a holo-analysis model can often be attributed to the performance of the control animals. The control animal performance probably acts as a proxy for 'undefined bird, environment and dietary factors' that influence the response to the feed additive. As noted in the models generated by Rosen, enzyme dosage was a less important term in describing the response to FCR gain in broilers than that of control bird performance or even sometimes other terms unrelated to the additive (Rosen, 2000; Rosen, 2002; Rosen, 2003). For example, the inclusion rate of fat in the diet had a larger effect than increasing enzyme dosages on broiler weight gain. This demonstrates that the additive or nutrient investigated is not always the most influential factor to describe animal performance in a data set, even though most if not all the initial trials were instigated to investigate the effect of the factor concerned. The advantage of such models is that the response to a given additive can be predicted when the other factors in the model are set to mimic the environment in which they might be used. For example, a commercial producer would want to consider the effect of the additive in diets and husbandry conditions that are relevant for their business.

8.11 Problems with Data Reporting in the Literature

Lack of consistency in the description of experiments is the greatest constraint when deriving a successful model from the literature (Kilkenny *et al.*, 2009). Researchers should therefore ensure that their papers report as much information relating to diet and husbandry as is possible. This would improve the repeatability of experiments and facilitate post hoc analysis. It is not foreseeable ahead of time which variables will be most important in describing the response to the variables investigated and thus the more data retained from each trial, the better the chance of teasing out these relationships later. Even with 500 lines of data lines available for analysis that have an output variable of interest, e.g. FCR, inconsistent experiment description can result in too few data available to construct a model, let alone a robust model. For example, age may be found to significantly influence FCR and be added to the model; however, it has only been reported in 80% of the data, reducing the available data for the model to 400 records. Lysine content may also be significant in the starter diet but also only be reported in 80% of the data. If the missing lines of age data lines do not coincide with the data lines for lysine, then only 300 of the original 500 data lines would remain with FCR, age and lysine being simultaneously reported. This is just one example and in fact analyses can be hampered even more than this in practice, especially when many independents are present in a model. This means the number of independent variables tested may have to be reduced to create a data set that is large enough to generate a robust model.

One example of the authors was a literature review containing 113 papers, which yielded over 1000 rows. However, this was reduced to less than 50 usable rows of data for the above reasons. This results in independent variables that may have a high significance from a biological viewpoint being excluded from models owing to the infrequency of their reporting.

8.12 Outputs and Interpretation

When it is possible to quantify the effects of changing the level of a variable in the diet on broiler carcass yield, for example, the optimum economic returns can be calculated based on the current prices of a variable (e.g. lysine) and carcass. This and more can be achieved with holo-analysis, where models can identify other contributory factors to a response. For example, the level of available phosphorous in the diet may alter the relationship between glycine level and body weight gain. There are often multiple high-cost variables present in these models (additives, amino acids, cereals, phosphorus, etc.) which can all be optimized on the basis of on price and level in the diet to optimize the output of interest whilst keeping costs to a minimum by manipulating the levels of variables in the diet based on changing prices, without needing a new model.

The power of holo-analysis is realized as the optimum diet is not always intuitive, especially when several interactions between independents are in the model. A model containing all the highest-cost nutrients and husbandry inputs would create the conditions for the greatest possible efficiency and cost-savings. Unfortunately, even inputs with high commercial relevance are often unreported in the literature and can be found insignificant within said models.

In the future, greater alignment between industry and academia would guarantee that both commercially relevant and costly inputs are always entered in academic research. Data are also widely collected in commercial settings at feed mills, during feed formulation, on breeder farms, broiler and layer farms. However, these data are often captured in isolation, which leaves vast potential for analysis undiscovered. With greater alignment of the breeder stock and mill and feed formulation, these data can be related to animal performance on farm and the holo-analysis process can be utilized on this extended data set, creating more robust and real-world applicable models. This will help a commercial company improve their own situation by generating bespoke models that take their geographical location, raw materials, feed milling processes, husbandry conditions, feed formulations and costs into account. There is a greater chance of maximizing profits within a company if the data used are their own. With data being collected in real-time, it would not take long for a medium- to large-sized company to produce enough data to create evolving models that are more robust than would be possible from the data currently available in the literature alone.

As mentioned previously, performance of control animals represents a significant portion of variation within most models. Within this variation are several unknown variables and by strengthening a data set with the data from commercial feed mills, broiler sheds and diet information, some of these unknowns may be teased out. Data that are often unreported in the literature are routinely captured by feed mills, for example, including throughput, energy consumed per tonne, conditioning time and temperature, pellet press temperature and temperature rise and cooling conditions amongst many other variables. Some of these conditions may well affect final profitability, through influence on the nutritional value of the diet or costs in diet manufacture. Holo-analysis could establish the optimum conditions needed to maximize profits during feed manufacture. The same may be true for husbandry conditions and diet ingredient selection and nutrient density. This is a realistic proposition and with complete and organized data sets; analysis could become a straightforward, routine practice for all sizes of outfits, allowing real-time optimization to keep up with the markets.

8.13 Conclusion

For these reasons, the applications of holo-analysis may be best utilized by commercial companies, rather than academia, who are producing large-scale datasets and can optimize animal production over the wide range of inputs required, from raw ingredients to feed manufacture and husbandry. This could then act as a catalyst for future research in academia, where the associations discovered during holo-analysis from the empirical data generated in the field can be further researched. There should also be more efforts by editors and referees of publications to establish a routine set of variables that must be included in all publications, such as complete feed formulations, mortality data and carcass characteristics, to allow greater data points and therefore lead to greater insights by subsequent holo-analysis. Above all else and regardless of where holo-analysis is utilized, it will be most successful if the tenet 'the quality of the data determines the quality of the output' is upheld.

References

Bedford, M. and O'Neill, H.M. (2016) Extending the value of the literature: Data requirements for holo-analysis and interpretation of outputs. In: *Nutrition Experiments in Pigs and Poultry: a Practical Guide.* CABI, Wallingford, UK, pp. 121–137.

Beker, A., Vanhooser, S., Swartzlander, J. and Teeter, R. (2004) Atmospheric ammonia concentration effects on broiler growth and performance. *Journal of Applied Poultry Research* 13(1), 5–9.

Choct, M. (2001) Enzyme supplementation of poultry diets based on viscous cereals. *Enzymes in Farm Animal Nutrition* 145–160.

Donaldson, W., Christensen, V., Garlich, J., McMurtry, J. and Olson, N. (1995) Exposure to excessive carbon dioxide: A risk factor for early poult mortality. *Journal of Applied Poultry Research* 4(3), 249–253.

Fletcher, J. (2014) Spurious correlations: Margarine linked to divorce. *BBC News Magazine* 26.

Glass, G.V. (1976) Primary, secondary, and meta-analysis of research. *Educational Researcher* 5(10), 3–8.

Hooge, D., Pierce, J., McBride, K. and Rigolin, P. (2010) Meta-analysis of broiler chicken trials using diets with or without Allzyme® SSF enzyme complex. *International Journal of Poultry Science* 9(9), 819–823.

Hooijmans, C.R., Leenaars, M. and Ritskes-Hoitinga, M. (2010) A gold standard publication checklist to improve the quality of animal studies, to fully integrate the Three Rs, and to make systematic reviews more feasible. *Alternatives to Laboratory Animals* 38(2), 167–182.

Kilkenny, C., Parsons, N., Kadyszewski, E., Festing, M.F., Cuthill, I.C. *et al.* (2009) Survey of the quality of experimental design, statistical analysis and reporting of research using animals. *PLoS ONE* 4(11), e7824.

Kilkenny, C., Browne, W., Cuthill, I.C., Emerson, M. and Altman, D.G. (2010) Animal research: reporting *in vivo* experiments: the ARRIVE guidelines. *British Journal of Pharmacology* 160(7), 1577.

Kilkenny, C., Browne, W.J., Cuthill, I.C., Emerson, M. and Altman, D.G. (2014) Improving bioscience research reporting: The arrive guidelines for reporting animal research. *Animals* 4(1), 35–44.

Letourneau-Montminy, M.P., Narcy, A., Lescoat, P., Bernier, J., Magnin, M. *et al.* (2010) Meta-analysis of phosphorus utilisation by broilers receiving corn-soyabean meal diets: influence of dietary calcium and microbial phytase. *Animal* 4(11), 1844–1853.

Marquardt, R.R. and Bedford, M. (2001) Future horizons. *Enzymes in Farm Animal Nutrition* 389–398.

Mehri, M. (2013) A comparison of neural network models, fuzzy logic, and multiple linear regression for prediction of hatchability. *Poultry Science* 92(4), 1138–1142.

Partridge, G.G. and Bedford, M.R. (2001) *Enzymes in farm animal nutrition*. CABI Publishing, Wallingford, UK.

Perai, A., Moghaddam, H.N., Asadpour, S., Bahrampour, J. and Mansoori, G. (2010) A comparison of artificial neural networks with other statistical approaches for the prediction of true metabolizable energy of meat and bone meal. *Poultry Science* 89(7), 1562–1568.

Reece, F. and Lott, B. (1980) Effect of carbon dioxide on broiler chicken performance. *Poultry Science* 59(11), 2400–2402.

Rosen, G. (2000) Multi-factorial assessment of exogenous enzymes in broiler pronutrition: Target and problems. *Proceedings of the 3rd European Symposium on Feed Enzymes*, Noordwijkerhout, The Netherlands.

Rosen, G. (2001a) Multi-factorial efficacy evaluation of alternatives to antimicrobials in pronutrition. *British Poultry Science* 42, S104–S105.

Rosen, G. (2001b) The nutritive value of phytase. *Proceedings of the 13th European Symposium on Poultry Nutrition,* World Poultry Science Association, The Netherlands.

Rosen, G. (2002) Microbial phytase in broiler nutrition. *Recent Advances in Animal Nutrition 2002*, 105–117.

Rosen, G. (2003) Setting and meeting standards for the efficient replacement of pronutrient antibiotics in broiler, turkey and pig nutrition. Proceedings of Gesunde Nutztiere. Heutiger

Stellewert der Futterzusatzstoffe in der Tierernaehrung. ETH-Zentrum, Zurich, Switzerland, pp. 72–88.

Rosen, G. (2004) Holo-analysis in animal nutrition. *Feed International* 25(12), 17–18.

Rosen, G. (2006) Holo-analysis. *Poultry Science* 85(6), 957–959.

Rosen, G.D. (1995) Antibacterials in poultry and pig nutrition. *Biotechnology in Animal Feeds and Animal Feeding* 172, 143.

Sales, J. (2014) Effects of access to pasture on performance, carcass composition, and meat quality in broilers: A meta-analysis. *Poultry Science* 93(6), 1523–1533.

Sauer, N., Emrich, K., Piepho, H.-P., Lemme, A., Redshaw, M. and Mosenthin, R. (2008) Meta-analysis of the relative efficiency of methionine-hydroxy-analogue-free-acid compared with DL-methionine in broilers using nonlinear mixed models. *Poultry Science* 87(10), 2023–2031.

Savegnago, R., Nunes, B., Caetano, S., Ferraudo, A., Schmidt, G. *et al.* (2011) Comparison of logistic and neural network models to fit to the egg production curve of White Leghorn hens. *Poultry Science* 90(3), 705–711.

Siegert, W., Ahmadi, H. and Rodehutscord, M. (2015) Meta-analysis of the influence of dietary glycine and serine, with consideration of methionine and cysteine, on growth and feed conversion of broilers. *Poultry Science* 94(8), 1853–1863.

Chapter 9
Mechanisms of Prebiotic Xylo-oligosaccharides on Improving Poultry Performance

Natalie Morgan*
Curtin University, Perth, Australia

Abstract

Arabinose substituted xylo-oligosaccharides (AXOS) and non-substituted xylo-oligosaccharides (XOS) are produced by hydrolytic degradation of arabinoxylans and xylans by endo-xylanases. These XOS are defined as prebiotics, based on their positive influence on the composition and activity of gastrointestinal microbiota, immunity and short-chain fatty acid production and ultimately bird performance. XOS are produced in the gastrointestinal tract in the presence of xylanase supplementation. There is also heightened interest in manufacturing XOS commercially and feeding them as a prebiotic supplement. Production of XOS both *in situ* and *in vitro* is influenced by several factors, including the susceptibility and accessibility of the xylan substrate, the specificity of the xylanase and the environmental conditions. This review examines the mechanisms behind how XOS are produced and how they act as prebiotics in poultry.

9.1 Introduction

Xylans are the most abundant hemicelluloses in the cell walls of cereals fed to poultry, with the non-starch polysaccharide (NSP) component in a typical broiler diet consisting of approximately 43% arabinoxylan (AX) for a maize–soybean-based diet and 47% for a wheat–soybean-based diet (Morgan, unpublished results). Xylanases are applied to poultry diets to eliminate the

*Corresponding author: natalie.morgan@curtin.edu.au

DOI: 10.1079/9781800622746.0009

antinutritional effects of xylan. Xylanase improves nutrient utilization and bird productive performance by reducing digesta viscosity and releasing encapsulated nutrients. In addition to these benefits, hydrolysis of xylan results in the production of small oligosaccharides that have been shown to possess prebiotic properties (Courtin *et al.*, 2008; Keerqin *et al.*, 2017; Morgan *et al.*, 2022). Endo-acting xylanases cleave the bonds of the linear polyxylose chains of AX and xylan, producing short-chain AXOS and XOS. The degree of polymerization and location of branching on the xylose backbone dictates the extent of viscosity effects and where xylanases can act to hydrolyse these long-chain molecules. The structures of the XOS generated vary in degree of polymerization (numbers of xylose residues in their backbone), degree of substitutions (ratio of arabinose to xylose), monomeric units and types of linkages. The number of xylose residues in the backbone can vary from two to ten. These xylose units are linked by β(1,4) bonds; for example, xylobiose (X_2;$C_{10}H_{18}O_9$), xylotriose (X_3;$C_{15}H_{26}O_{13}$) and xylotetraose (X_4;$C_{20}H_{34}O_{17}$). The size of these oligosaccharides determines which species in the microbiota can utilize them as a substrate and thus the product generated from them. XOS can also be linked to substituents such as acetyl groups and uronic acids (Caparros *et al.*, 2007). Increased understanding of microbiota response to different structures and sizes of XOS is required to develop xylanases and prebiotic additives with increased efficacy.

The aim when applying xylanases is to produce XOS, as opposed to complete hydrolysis into monomers. D-Xylose and L-arabinose are poorly metabolized by monogastric animals and often increase the level of moisture in droppings and severity of diarrhoea. This is due to the osmotic properties of the unabsorbed pentose sugars stimulating microbial activity in the intestinal tract, resulting in inflow of water into the intestinal lumen (Zyla *et al.*, 1999). Thus, xylanases applied to animal diets are intended to partially hydrolyse xylan into oligomers and smaller polymers, rather than into monomers. There are several factors that influence XOS production in the presence of xylanase, including the susceptibility and availability of the xylan substrate, the specificity and mechanisms of the enzyme and the environmental conditions of the reaction. This review examines these factors, along with the effect of the resulting XOS on the composition and activity of the gastrointestinal microbiota, short-chain fatty acid (SCFA) production and immunity and, ultimately, bird performance.

9.2 Factors Influencing XOS Production

9.2.1 Ingredient variability

The concentrations and structural features of AX and xylan differ widely between different ingredients and batches of the same feed ingredient. The quantity and sequence of distribution of the structural elements in AX are often characterized by measuring the arabinose to xylose ratio. The distribution of arabinose along the xylose backbone is non-random, but the structure is more regular for some ingredients than for others. Therefore, even in cereals where the arabinose:xylose ratio is similar, the distribution of arabinose on the xylan backbone can vary markedly, such as seen in wheat and rye (Izydorczyk and Biliaderis, 1995). The structure of AX is also different in the endosperm compared with the aleurone and bran. This highlights that it is not possible to predict XOS production in the gastrointestinal tract based on the quantity of AX or xylan in the ingredient being fed, as several factors determine 'accessibility' and 'susceptibility' of this ingredient to enzyme effects. A primary contributor is growth conditions. Morgan *et al.* (2020) measured XOS production from different wheat and barley samples from different states across Australia, in either the presence or absence of xylanase, under *in vitro* conditions that mimicked the gizzard and ileum. The results showed that XOS production was incredibly diverse between different batches of the same ingredient, especially in the presence of xylanase, due to variation in growth conditions and the impact of this on chemical composition. In this study it appeared that samples grown in warmer and drier conditions were more susceptible to the effects of xylanase. These results agree with Olukosi and Bedford (2019) and Amerah *et al.* (2009), who found response of wheat to xylanase varies greatly depending on the endosperm structure of wheat and wheat hardness. Xylans are either water soluble or insoluble; those containing branching linkages, such as arabinose and glucuronic acid, are more soluble than those that are not, because they have a lower capacity to bind to cellulose. Less branched xylans can bind tightly to cellulose and lignin, limiting the ability of the plant cell wall to be broken down (Mortimer *et al.*, 2010). This suggests that to successfully predict xylanase efficacy it is imperative to understand the solubility of the xylan in the dietary ingredients and how this will influence XOS production and the size of the XOS fractions.

9.2.2 Xylanase family

The heterogeneity and complexity of xylanase has resulted in a range of xylanases with different specificities, meaning xylanases cannot be classified based on the substrate alone. Instead, xylanases are classified into enzyme families based on their catalytic modules. Xylanases used in poultry diets are usually grouped into glycoside hydrolase (GH) families 10 and 11, based on amino acid sequence similarity. Family 10 and family 11 xylanases differ from each other with regards to their specificities, with family 10 resulting in comparatively greater production of XOS than family 11 (Biely *et al.*, 1997). Family 11 xylanases only hydrolyse xylan and principally only on the xylose unit at the centre of the oligosaccharide chain. Family 10, however, prefer to act on groups at the termini on the xylan bonds, particularly the reducing end, have lower substrate specificity and can work on any location with a side branch. Consequently, family 11 are generally most active on longer chain XOS, whereas family 10 work more on smaller chain XOS and have a more flexible structure. Family 10 can also hydrolyse highly substituted xylans and short XOS more effectively, resulting in greater adsorption and hydrolytic efficiency on insoluble xylans, because they can continuously hydrolyse bonds along the xylan chain and at any positions where there are branches. Family 11 xylanases, however, have greater effects on soluble xylan as they act on the centre of xylan chains. Morgan *et al.* (2017) subjected pure wheat xylan to both family 10 and family 11 xylanase *in vitro* and observed that family 10 xylanase had comparatively greater hydrolytic efficiency than family 11, as determined by the quantity of xylo-oligosaccharides produced. This was probably because family 10 could cleave more linkages closer to the substituent groups. This illustrates that it is important to consider the enzyme family when seeking to produce xylo-oligosaccharides, both in situ in the animal and ex situ when manufacturing XOS feed additives. Unfortunately, failure to accurately report enzyme assay results and enzyme classes within the literature currently makes it difficult to interpret the efficacy of xylanases across different research studies.

9.2.3 Environmental conditions

Environmental conditions in the gastrointestinal tract dictate XOS production, specifically pH, temperature and the amount of time the xylan is exposed to xylanase. Most xylanases used in poultry feeds are produced by genetically modified production strains of the fungus *Trichoderma reesei*; these are alkaline enzymes but are relatively pH stable, due to the moderate balance of basic and acid residues. Morgan *et al.* (2017) observed that pH had a significant impact on XOS production, influencing the structure of both xylan and xylanase. In this study, conversion of pure xylan into XOS was greatest at pH 2.5 compared to any other pH ranging from 2 to 7, with approximately 25% of the total xylan being manufactured into X_2–X_4 at this pH. This was followed by pH 3 and 2. This is an interesting finding because it was predicted that the optimum pH would be 5.5, as found in ileal conditions. This suggests xylan degradation may occur primarily at the beginning of the digestive tract. Courtin *et al.* (2009) observed similar, with optimum hydrolysis of oligosaccharide linkages taking place at pH 2 and 3. Also, Vazquez *et al.* (2000) states that acidic conditions are needed to successfully break down xylan obtained from alkali extraction. Morgan et al. (2017) also presented an interaction between pH and the amount of time the xylan is exposed to xylanase; at pH 2.5 more XOS was produced when measured at 60 minutes compared to 120 minutes but at pH 7 the opposite was true, with greater manufacture after 120 minutes. This agrees with Kumar and Wyman (2008) who observed that degradation of xylan is sped up in acidic conditions due to direct acidic hydrolysis and reduction in the role of side reactions. This suggests that a longer timeframe is needed to degrade xylan at a higher pH, implying that it may be advantageous to reduce digesta transit rate along the gastrointestinal tract to produce more prebiotic XOS. This is also highlighted by Aguedo *et al.* (2015), who found that oligosaccharide release from destarched wheat bran exposed to xylanase was fast during the first hour, but above 4 hours it was slow and poorly significant.

Temperature may also influence XOS production, with higher temperatures potentially resulting in more XOS production and reduced time required to convert xylan into XOS. However, if temperature is too high xylanase will begin to denature faster than it can hydrolyse xylan. For example, He *et al.* (2009) and La Grange *et al.* (1996) observed highest activity of a *T. reesei* xylanase at 60°C, but the enzyme was not stable at this temperature and activity decreased dramatically over time. It must be noted that the source of the xylan and xylanase dictates the influence of environmental conditions on XOS production. For example, Kambourova *et al.* (2007) found that a longer incubation time resulted in heightened xylose production when xylan from beechwood was exposed to xylanases, but reduced xylose production with xylan from oat spelt. *In vitro* assays that mimic the environmental conditions of the poultry gastrointestinal tract have the potential to act as sensitive and cost-effective tools for the evaluation of both xylanase efficacy and XOS manufacture from dietary ingredients. This was investigated by Morgan *et al.* (2020) by directly comparing the quantity of X_5–X_2 measured in the gizzard and ileum of broiler chickens to that produced when exposing the same dietary treatments to a two-step *in vitro* digestion assay that mimicked the pH, temperature and time period of the gastric and small intestine phases; relatedness of the two methods was strong, illustrating that this *in vitro* assay was successful at predicting xylanase effect on XOS production.

9.3 Prebiotic Effects of Xylo-oligosaccharides

XOS are not significantly degraded by gastric acid or digestive enzymes and are not absorbed by intestinal mucosa (Imaizumi *et al.*, 1991). This means that they reach the distal parts of the intestinal tract intact and are selectively fermented by the gastrointestinal microbiota, particularly probiotic bacteria, which produce short-chain fatty acids. This induces positive impacts on the composition and activity of gastrointestinal microbiota. Consequently, XOS can be defined as prebiotics (Lecerf *et al.*, 2012). These prebiotic properties are dependent on the impact that enzymatic treatment has on degrees of polymerization and substitution. The potential prebiotic effects of XOS include optimizing colon function, increasing or changing composition of SCFAs, increasing mineral absorption, immune stimulation and increased ileal villus length. The beneficial technological features of XOS include stability at acidic pH, heat resistance, ability to achieve significant biological effects at low daily doses, low calorie content and no toxicity (Carvalho *et al.*, 2013). XOS can be derived from a wide range of natural sources, including plants and various agricultural by-products, such as bamboo shoots, fruits, vegetables, milk and honey. Of particular interest is the possibility to produce XOS from agricultural residues, as they are cheap and abundantly available (Samanta *et al.*, 2017); examples include cotton, tobacco and sunflower stalks (Akpinar *et al.*, 2009), sugarcane bagasse (Peng *et al.*, 2009) and maize stalk (Ergues *et al.*, 2012), as well as lignocellulosic biomass (Aachary and Prapulla, 2011). Morgan *et al.* (2018) examined the effects of AXOS derived from a wheat starch-milling by-product on net energy and nutrient utilization in broilers and observed improvements in performance, energy utilization, nutrient digestibility, microbiota composition and SCFA production. This study highlights the multi-faceted impact of XOS in the poultry gastrointestinal tract.

9.3.1 Effect of XOS on microbiota composition

The microbiome found early in the gastrointestinal tract, from the crop to the small intestine, utilize some of the soluble polysaccharides and fermentable oligosaccharides, leaving behind the less readily digestible polysaccharides such as β-glucans and AX. These then enter the distal ileum, caeca and colon, where the microbiome must adapt to ferment and utilize them. These 'fermentation chambers' are dominated mainly by species belonging to the phyla *Firmicutes* (up to 75%) and *Bacterioidetes* (between 10% and 50%) (Dumonceaux *et al.*, 2006; Gong *et al.*, 2002; Lu *et al.*, 2003), although most bacteria species in the gastrointestinal tract remain unidentified. XOS has a bifidogenic effect, maintaining balance of these bacteria and acting as a carbon and energy source, promoting the growth of beneficial bacteria such as *Bifidobacterium adolescentis*, *B. longum*, *Lactobacillus brevis* and *L. fermentum* (Moura *et al.*, 2007; Pourabedin *et al.*, 2015). According to Aachary and Prapulla (2011), *Bifidobacteria* tend to prefer short, unsubstituted XOS and are not able to utilize double-substituted AXOS. Rycroft *et al.* (2001) also showed that XOS stimulates an increase in *Bifidobacteria* levels to a greater extent compared to other oligosaccharides and Mäkeläinen *et al.* (2010), Zeng *et al.* (2007) and Samanta *et al.* (2017) reported that XOS were fermented with high specificity by strains of *Bifidobacteria*. Moreover, Ding *et al.* (2018) found that the number of *Bifidobacteria* in the caecum in White Lohmann laying hens increased linearly with an increasing level of XOS, ranging from 0 to 0.05%.

This suggests that the increase in beneficial bacteria in the caeca makes these the dominant bacteria, meaning they can competitively inhibit the growth of harmful bacteria, such as *Escherichia coli*, as has been observed when feeding broilers mannan-oligosaccharides (Baurhoo *et al.*, 2007). Colonization of bacteria on mucosal tissue is a vital step for inducing infection and to colonize the mucosal surface bacteria must first bind to the epithelial cells (Jahanian and Ashnagar, 2015). XOS possibly competitively binds to pathogens, preventing them from adhering to the host's epithelium and inducing disease (Chen *et al.*, 2021).

If there is a greater abundance of protein fermentation compared to fibre fermentation in the ileum and caeca then putrefaction issues may occur. Also, when diets are changed between phases, this puts a great deal of pressure on the microbiome to quickly adapt to the different nutrient availability (Ijaz *et al.*, 2018), which can lead to heightened quantities of pathogenic bacteria being present, particularly if the stream of protein has been increased. Establishing the fibre-degrading microbiome as early as possible could help ensure the transition from using starch, to protein, to fibre as fuel could be much more controlled. Thus, providing a stable supply of fermentable substrates, such as XOS, to diets could guarantee that there is sufficient fuel for the beneficial microbiota regardless of diet composition.

9.3.2 Effect of XOS on short-chain fatty acid production

XOS act as an energy source for saccharolytic bacteria, such as *Bifidobacter* and *Lactobacillus spp.*, in the caeca. This benefits the bird by increasing energy utilization from the diet in the form of SCFAs, namely butyrate, acetate and lactate. This has been confirmed in several *in vitro* and *in vivo* studies (Broekaert *et al.*, 2011; Scott *et al.*, 2014). For example, Morgan *et al.* (2018) observed that total SCFA concentration, along with lactic and formic acid concentration, was higher in the ileum of birds fed diets supplemented with 2% AXOS compared to those fed 2% AX. Lactate production usually occurs because of saccharide and starch fermentation, whereas fermentation of less readily fermentable fibres results in acetate and butyrate production (Mahmood and Guo, 2019). Lactate generated in the ileum, if not absorbed and used as an energy source, will enter the caeca where it will be fermented further into butyrate. This is due to cross-linking between lactate-producing bacteria and lactate-utilizing butyrate-producing bacteria from *Clostridium* cluster XIV (Duncan *et al.*, 2004). This metabolic cross-feeding can help to stabilize luminal pH and increase mineral solubility. Butyrate itself fuels epithelial cells in the colon, exerts anti-inflammatory activities and increases intestinal epithelial integrity, which results in improved growth performance and positive changes to intestinal microbiota composition and metabolic activity. For example, De Maesschalck *et al.* (2015) observed that feeding broilers XOS derived from maize cob increased the number of gene copies encoding the enzyme for butyrate production, as well as causing increased feed conversion, villus length in the ileum and number of lactobacilli in the colon and *Clostridium* cluster XIVa in the caeca. Butyrate-producing bacteria and lactobacilli were also comparatively more abundant in the caecum and colon of birds fed XOS than those fed the control diet. A number of other studies, including Canani *et al.* (2011) and Guilloteau *et al.* (2010), have also observed that butyrate has a beneficial effect on gastrointestinal function, through anti-inflammatory properties, increasing intestinal epithelial integrity and as a fuel for epithelial cells. The consequence of this is positive effects on growth performance and changes to microbiota composition and metabolic activity in the intestine.

As well as directly initiating SCFA production, XOS production may also potentially be 'signalling' the microbiome to further degrade fibre present in the digesta, resulting in heightened SCFA production. This theory could possibly explain why there are consistently such high levels of SCFA produced relative to the small amount of XOS present. This has been illustrated by Bedford and Apajalahti (2018) who observed that birds reared on xylanase had a caecal microbiome that was much better adapted to digesting xylan compared to those that had never been fed xylanase. This result suggests that xylanase presence stimulates bacteria in the ileum and caeca to produce their own fibre-degrading enzymes, through signalling molecules. This would suggest that the heightened SCFA production observed in the presence of XOS is more likely a consequence of its ability to stimulate the microbiome to produce more enzymes than its capacity as a prebiotic. This was highlighted by Morgan *et al.* (2022) who observed increased caeca SCFA concentration and xylanase and cellulase activity from supplementing 2000 mg/kg XOS and fermentable fibre into sorghum based diets fed to broilers.

9.3.3 Effect of XOS on immunity and metabolism

A healthy microbiome has extensive immunogenicity (Wickramasuriya *et al.*, 2022). For example, lactobacilli can adhere to the mucosal layer and epithelium, causing immunomodulation and offering protection to the intestinal barrier by antagonistic activities against pathogens (Servin, 2004). This guarantees for a minimally activated immune system and thus reduced inflammatory response. Additionally, SCFAs enhance intestinal, non-specific immune-modulatory mechanisms by stimulating the expression of mucin glycoprotein in intestinal epithelial cells to combat pathogens (Willemsen *et al.*, 2003). This highlights that the impact of XOS on SCFA production and microbiota composition may potentially cause enhanced immunity in poultry. This was illustrated by Yuan *et al.* (2018) who found that supplementing 2 mg/kg XOS to broilers fed maize–soybean-based diets resulted in an increase in plasma IgA and IgG and down-regulated mRNA expression of lipopolysaccharide (LPS)-induced TNF-alpha factor, Toll-like receptor-5 and interferon gamma in the jejunum at day 21. They also observed improved average daily gain and increased acetate and butyrate concentration in the caeca, suggesting that XOS reduced the expression of cytokine genes and improved broiler immune function by stimulating SCFA production. Niewold *et al.* (2012) also confirmed that XOS directly influenced the immune system by binding to toll-like receptors and Sun *et al.* (2013) showed that XOS increased the titre of antigens, resulting in improved cellular and humoral immunity.

There is evidence to suggest that AXOS has an impact on prevalence of *Salmonella*. For example, Eeckhaut *et al.* (2008) observed that feeding 0.4% AXOS to broilers orally inoculated with 2.5×10^9 CFU of *Salmonella* Enteritidis resulted in a significant reduction in colonization of *Salmonella* in the caeca and translocation of *Salmonella* to the spleen compared to birds fed the control diet. Furthermore, Keerqin *et al.* (2017) observed that AXOS displayed preventative actions against necrotic enteritis, illustrated by a reduced decline in bird performance and severity of intestinal lesions. This suggests that AXOS has a protective effect against the colonization of *Salmonella* and *Clostridium perfringens*. This may in part be due to the increased number of bifidobacteria in the caeca physically decreasing the count of these pathogens, or through increased butyric acid production; butyric acid decreases invasion of *Salmonella* and *Clostridum perfringens* on gut epithelial cells and colonization of the caeca and internal organs, potentially mediated by a specific effect on invasion gene expression (Van Immerseel *et al.*, 2005). Also, enhanced butyrate production because of XOS can increase defence mechanisms against pathogens, by stimulating mucin glycoprotein expression in intestinal epithelial cells (Timbermont *et al.*, 2010). Another interesting observation from the study conducted by Eeckhaut *et al.* (2008) was that it revealed that degree of polymerization plays a substantial role in the immune effects of XOS, presenting that there was less *Salmonella* colonization when feeding AXOS with an average degree of polymerization of nine compared to three.

XOS has also been shown to have a positive impact on thyroid function. This was illustrated by Zhenping *et al.* (2013) who found that the levels of triiodothyronine, thyroxine and insulin in the blood and titres of antibody to avian influenza H5N1 virus were higher in broilers fed XOS produced from straw chaff compared to those fed the control diet. It appears that it specifically influences the hormones involved in growth, based on its influence on triiodothyronine, which has previously been shown to relate positively to growth (Chang *et al.*, 2003) and improved body weight gain and feed conversion in birds fed 10 g XOS/kg. Thus, it seems that XOS plays a diverse role in enhancing immunity, primarily through its influence on microbiota composition and SCFA production. Moreover, Samanta *et al.* (2017) found that feeding broilers 0.5% XOS from maize husks resulted in significantly reduced cholesterol, very low-density lipoprotein and glucose concentration, as well as numerically lower triglycerides. This effect on cholesterol and glucose may be due to a higher proportion of SCFA from XOS fermentation, namely propionate which is reported to inhibit the metabolic pathways of gluconeogenesis and cholesterol synthesis. Li *et al.* (2017) observed similar in layers, in which it was found that plasma glutamic-pyruvic transamine, cholesterol, high-density lipoprotein and very low-density lipoprotein decreased linearly with an increasing level of XOS, from 0 to 0.5 g/kg.

9.3.4 Effect of XOS on bird performance

Current published data present supplemented XOS doses ranging from 0.002 g/kg (Yuan *et al.*, 2018) to 0.025–0.1 g/kg (Suo *et al.*, 2015; Singh *et al.*, 2021), 1–5 g/kg (Craig *et al.*, 2020; Courtin *et al.*, 2008; Bautil *et al.*, 2020; Samanta *et al.*, 2017; De Maesschalck *et al.*, 2015; Morgan *et al.*, 2022) and 10–20 g/kg (Zhenping *et al.*, 2013). Variable responses in bird performance have been observed across these studies. For example, when supplementing 5 g/kg XOS to wheat-based diets, De Maesschalck *et al.* (2015) and Courtin *et al.* (2008) observed improvements in feed conversion ratio (FCR) but Bautil *et al.* (2020) saw no effect on bird performance. Additionally, Craig *et al.* (2020) found XOS supplementation had a negative impact on performance compared to feeding xylanase. The variability observed in performance responses between studies is probably due to differences in content and physiochemical properties of the dietary xylan offered to the birds, dictating the fermentation capacity of the microbiota. Bedford and Apajalahti (2018) suggested it is fermentation of fibre rather than direct manufacture of oligosaccharides that provides additional energy and feedback to the gizzard, thus dictating performance. In maize-based diets, XOS supplementation at 0.1 g/kg had a positive impact on FCR in the study by Suo *et al.* (2015) but not in the study by Singh *et al.* (2021). However, Yuan *et al.* (2018) saw that just 0.002 g/kg XOS was sufficient to induce positive effects on weight gain. In contrast, Zhenping *et al.* (2013) observed improvements on FCR with a high level of 10 g/kg XOS but not 5 g/kg. The positive effects observed in birds fed maize-based diets highlight the benefits of providing XOS to birds fed NSP-poor diets, as fuel for beneficial microbiota, to upregulate xylan fermentation pathways and stimulate proliferation of xylan-degrading bacteria. This was also illustrated by Morgan *et al.* (2022), in which supplementing 2 g/kg XOS improved body weight gain in birds fed sorghum-based diets, a consequence of heightened SCFA and endogenous enzyme activity in the caeca.

In layers, performance responses to XOS are also inconsistent. Zhou *et al.* (2009) supplemented 0.15 and 0.2 g/kg XOS into layer diets and found it elevated mean daily laying rate and egg weight, and at 0.2 g/kg it increased eggshell intensity, eggshell thickness and egg rate. Zhou *et al.* (2021) also observed improved FCR when feeding 0.2 and 0.4 g/kg XOS. Conversely, Li *et al.* (2017) showed that XOS had no effect on production performance but significantly improved eggshell thickness and eggshell relative weight. This may be attributed to the positive effect of XOS on calcium digestibility; it is likely that XOS influences mineral absorption rate, namely Ca^{2+} and Mg^{2+}, by altering the metabolic activity of the beneficial bacterial colony in the intestine (Roberfroid, 2000) or increasing the expression of Ca-binding proteins (Swiatkiewicz *et al.*, 2010). Also, Ding *et al.* (2018) found that XOS increased $1,25(OH)_2D_3$ measured in the plasma, which may partly explain this result because $1,25(OH)_2D_3$ promotes absorption of calcium in intestinal epithelial cells. These presented studies observed effects of XOS in maize-based diets, but there is a deficit of studies presenting the effects in wheat-based diets. Further research is warranted into the ability of XOS to improve egg production and quality.

XOS has been shown to have a positive impact on intestinal morphology and the gastrointestinal environment. For example, De Maesschalck *et al.* (2015) observed longer villi in birds fed XOS compared to those fed the control diet and Suo *et al.* (2015) found that crypt depth in the duodenum was significantly lower in birds fed 75 mg/kg XOS compared to birds fed lower levels. Furthermore, Ding *et al.* (2018) found that, in layers, villus height and the villus-to-crypt ratio of the jejunum, as well as the relative length of the jejunum and villus height in the duodenum, was increased linearly with an increasing level of dietary XOS. This positive effect on intestinal morphology may be partly due to butyrate production, as butyrate stimulates glucagon-like peptide 2 production, which is a hormone shown to have positive effects on growth performance, intestinal morphology, villus height and crypt cell proliferation (Liu *et al.*, 2021). Alternatively, it may be related to an increase in the absorption of nutrients and the positive effect of butyrate on intestinal morphology (Abdelqader and Al-Fataftah, 2016; Hu and Guo, 2007; Sikander *et al.*, 2017).

There are potentially several advantages to supplementing XOS in combination with xylanase, compared to just xylanase alone. XOS are more thermostable than xylanases, making it possible to include XOS before feed pelleting (Courtin *et al.*, 2009). Also, efficacy of xylanase is dependent on composition of the feed, whereas XOS is effective in all diets, not just those containing wheat, barley and rye, which is important given the wide use of maize in a number of countries. XOS are also not impacted by the variable levels of endogenous xylanase inhibitors in wheat. Courtin *et al.* (2008) stated that the positive effects of supplementing AXOS on feed utilization efficiency was similar to that obtained by adding xylanase to a wheat-based diet. In this study they observed that xylanase, unlike supplemental AXOS, did not cause a significant increase in intestinal *Bifidobacteria* spp., which shows AXOS and xylanase work by different modes of action. Morgan *et al.* (2018) found similar, in that feeding AXOS manufacture *in vitro* was more efficient than feeding birds arabinoxylan and xylanase in combination. Furthermore, Craig *et al.* (2020) observed that supplementation with XOS increased the concentration of SCFA in the caeca, significantly more so compared to feeding xylanase alone. The benefits of supplementing XOS and xylanase in combination on FCR have been presented by González-Ortiz *et al.* (2021) and Morgan *et al.* (2021).

9.4 Conclusion

XOS as prebiotics have received heightened interest recently, owing to their ability to enhance bird performance through their influence on microbiota composition, SCFA production and immunity, potentially providing an alternative to in-feed antibiotics. Additionally, XOS can be manufactured from waste and by-products, meaning they are cost effective and sustainable. Further research is warranted to ascertain more in-depth understanding of the mechanisms behind how XOS manipulate the microbiota but evidence thus far suggests they are promising prebiotics.

References

Aachary, A.A. and Prapulla, S.G. (2011) Xylooligosaccharides (XOS) as an emerging prebiotic: microbial synthesis, utilization, structural characterization, bioactive properties and applications. *CRFSFS* 10, 2–16.

Abdelqader, A. and Al-Fataftah, A.R. (2016) Effect of dietary butyric acid on performance, intestinal morphology, microflora composition and intestinal recovery of heat-stressed broilers. *Livestock Science* 183, 78–83.

Aguedo, M., Ruiz, H.A. and Richel, A. (2015) Non-alkaline solubilization of arabinoxylans from destarched wheat bran using hydrothermal microwave processing and comparison with the hydrolysis by an endoxylanase. *Chemical Engineering and Processing: Process Intensification* 96, 72–82.

Akpinar, O., Erdogan, K. and Bostanci, S. (2009) Enzymatic production of xylooligosaccharides from selected agricultural wastes. *Feed Bioproducts and Processing* 87, 145–151.

Amerah, A., Ravindran, V. and Lentle, R.G. (2009) Influence of wheat hardness and xylanase supplementation on theperformance, energy utilisation, digestive tract develop-ment and digesta parameters of broiler starters. *Animal Production Science* 49, 71–78.

Baurhoo, B., Phillip, L. and Ruiz-Feria, C.A. (2007) Effects of purified lignin and mannan oligosaccharides on intestinal integrity and microbial populations in the ceca and litter of broiler chickens. *Poultry Science* 86, 1070–1078.

Bautil, A., Verspreet, J., Buyse, J., Goos, P., Bedford, M. and Courtin, C. (2020) Arabinoxylan-oligosaccharides kick-start arabinoxylan digestion in the aging broiler. *Poultry Science* 99, 2555–2565.

Bedford, M.R. and Apajalahti, J. (2018) Exposure of a broiler to a xylanase for 35d increases the capacity of cecal microbiome to ferment soluble xylan. *Poultry Science* 97, 98–99.

Biely, P., Vrsanská, M., Tenkanen, M. and Kluepfel, D. (1997) Endo-beta-1, 4-xylanase families: Differences in catalytic properties. *Journal of Biotechnology* 57, 151–166.

Broekaert, W.F., Courtin, C.M., Verbeke, K., Van de Wiele, T., Verstraete, W. and Delcour, J.A. (2011) Prebiotic and other health-related effects of cerealderived arabinoxylans, arabinoxylan-oligosaccharides, and xylooligosaccharides. *Critical Reviews in Food Science and Nutrition* 51,178–194.

Canani, R.B., Costanzo, M.D., Leone, L., Pedata, M., Meli, R. and Calignano, A. (2011) Potential beneficial effects of butyrate in intestinal and extraintestinal diseases. *World Journal of Gastroenterology* 17, 1519–1528.

Caparros S., Garrote, G., Ariza, J., Diaaz, M.J.S. and Loapez, F. (2007) Xylooligosaccharides production from *Arundo donax*. *Journal of Agricultural and Food Chemistry*. 55, 5536–5543.

Carvalho, A.F.A., Oliva Neto, P.D., Silva, D.F.D. and Pastore, G.M. (2013) Xylo oligosaccharides from lignocellulosic materials: Chemical structure, health benefits and production by chemical and enzymatic hydrolysis. *Food Research International* 5175–5185.

Chang, S.C., Lin, M.J., Croom, J. and Fan, Y.K. (2003) Administration of triiodothyronine and dopamine to broiler chicks increases growth, feed conversion and visceral organ mass. *Poultry Science* 82, 285–293.

Chen, Y., Xie, Y., Zhin, R., Liu, L., Lin, C. *et al.* (2021) Effects of xylo-oligosaccharides on growth and gut microbiota as potential replacements for antibiotics in weaning piglets. *Frontiers in Microbiology* 12, 641172.

Courtin, C.M., Broekaert, W.F., Swennen, K., Lescroart, O., Onagbesan, O. *et al.* (2008) Dietary inclusion of wheat bran arabinoxylooligosaccharides induces beneficial nutritional effects in chickens. *Cereal Chemistry* 85, 607–613.

Courtin, C.M., Swennen, K., Verjans, P. and Delcour, J.A. (2009) Heat and pH stability of prebiotic arabinoxylooligosaccharides, xylooligosaccharides and fructooligosaccharides. *Food Chemistry* 112, 831–837.

Craig, A., Khattak, F., Hastie, P., Bedford, M.R. and Olukosi, O. (2020) Xylanase and xylo-oligosaccharide prebiotic improve the growth performance and concentration of potentially prebiotic oligosaccharides in the ileum of broiler chickens. *British Poultry Science* 61 (1), 70–78.

De Maesschalck, C.D., Eeckhaut, V., Maertens, L., De Lange, L., Marchal, L. *et al.* (2015) Effects of xylo-oligosaccharides on broiler chicken performance and microbiota. *Applied Environmental Microbiology* 81, 5880–5888.

Ding, X.M., Li, D.D., Bai, S.P., Wang, J.P., Zeng, Q.F. *et al.* (2018) Effect of dietary xylooligosaccharides on intestinal characteristics, gut microbiota, cecal short-chain fatty acids, and plasma immune parameters of laying hens. *Poultry Science* 97, 874–881.

Dumonceaux, T.J., Hill, J.E., Hemmingsen, S.M. and Van Kessel, A.G. (2006) Characterization of intestinal microbiota and response to dietary virginiamycin supplementation in the broiler chicken. *Applied Environmental Microbiology* 72, 2815–2823.

Duncan, S.H., Louis, P. and Flint, H.J. (2004) Lactate-utilizing bacteria, isolated from human feces that produce butyrate as a major fermentation product. *Applied Environmental Microbiology* 70, 5810–5817.

Eeckhaut, V. Van Immerseel, F. Dewulf, J. Pasmans, F. Haesebrouck, F. *et al.* (2008) Arabinoxylooligosaccharides from wheat bran inhibit *Salmonella* colonization in broiler chickens. *Poultry Science* 87, 2329–2334.

Ergues, I., Sanchez, C., Mondragon, I. and Labidi, J. (2012) Effect of alkaline and autohydrolysis processes on the purity of obtained hemicellulose from corn stalks. *Bioresource Technology* 103, 239–248.

Gong, J., Forster, R.J., Yu, H., Chambers, J.R., Sabour, P.M. *et al.* (2002) Diversity and phylogenetic analysis of bacteria in the mucosa of chicken ceca and comparison with bacteria in the cecal lumen. *FEMS Microbiology Letters* 208, 1–7.

González-Ortiz, G., dos Santos, T.T. and Bedford, M.R. (2021) Evaluation of xylanase and a fermentable xylo-oligosaccharide on performance and ileal digestibility of broiler chickens fed energy and amino acid deficient diets. *Animal Nutrition* 7, 488–495.

Guilloteau, P., Martin, L., Eeckhaut, V., Ducatelle, R., Zabielski, R. and Van Immerseel, F. (2010) From the gut to the peripheral tissues: the multiple effects of butyrate. *Nutrition Research Reviews* 23, 366–384.

He, J., Yu, B., Zhang, K., Ding, X. and Chen, D. (2009) Expression of endo-1,4-beta-xylanase from *Trichoderma reesei* in Pichia pastoris and functional characterization of the produced enzyme. *BMC Biotechnology* 9, 56.

Hu, Z. and Guo, Y. (2007) Effects of dietary sodium butyrate supplementation on the intestinal morphological structure, absorptive function and gut flora in chickens. *Animal Feed Science and Technology* 132, 240–249.

Ijaz, U.Z., Sivaloganathan, L., McKenna, A., Richmond, A., Kelly, C. *et al.* (2018) Comprehensive longitudinal microbiome analysis of the chicken cecum reveals a shift from competitive to environmental drivers and a window of opportunity for campylobacter. *Frontiers in Microbiology* 9, 2452.

Imaizumi, K., Nakatsu, Y., Sato, M., Sedarnawati, Y. and Sugano. M. (1991) Effects of xylooligosaccharides on blood glucose, serum and liver lipids and cecum short-chain fatty acids in diabetic rats. *Agricultural and Biological Chemistry* 55, 199–205.

Izydorczyk, M.S. and Biliaderis, C.G. (1995) Cereal arabinoxylans: Advances in structure and physicochemical properties. *Carbohydrate Polymers* 28, 33–48.

Jahanian, R. and Ashnagar, M. (2015) Effect of dietary supplementation of mannan-oligosaccharides on performance, blood metabolites, ileal nutrient digestibility, and gut microflora in *Escherichia* coli-challenged laying hens. *Poultry Science* 94, 2165–2172.

Kambourova, M., Mandeva, R., Fiume, I., Maurelli, L., Rossi, M. and Morana, A. (2007) Hydrolysis of xylan at high temperature by co-action of the xylanase from *Anoxybacillus flavithermus* BC and the beta-xylosidase/alpha-arabinosidase from *Sulfolobus solfataricus Oalpha*. *Journal of Applied Microbiology* 102, 1586–1593.

Keerqin, C., Morgan, N.K., Wu, S., Swick, R.A. and Choct, M. (2017) Dietary inclusion of arabinoxylo-oligosaccharides in response to broilers challenged with subclinical necrotic enteritis. *British Poultry Science* 58, 418–424.

Kumar, R. and Wyman, C.E. (2008) The impact of dilute sulfuric acid on the selectivity of xylooligomer depolymerization to monomers. *Carbohydrate Research* 343, 290–300.

La Grange, D.C., Pretorius, I.S. and Van Zyl, W.H. (1996) Expression of a *Trichoderma reesei* β-Xylanase Gene (XYN2) in *Saccharomyces cerevisiae. Applied and Environmental Microbiology* 1036–1044.

Lecerf, J.-M., Dépeint, F., Clerc, E., Dugenet, Y., Niamba, C.M. *et al.* (2012) Xylo-oligosaccharide (XOS) in combination with inulin modulates both the intestinal environment and immune status in healthy subjects, while XOS alone only shows prebiotic properties. *British Journal of Nutrition* 108, 1847–1858.

Li, D.D., Ding, X.M., Zhang, K.Y., Bai, S.P., Wang, J.P. *et al.* (2017) Effects of dietary xylooligosaccharides on the performance, egg quality, nutrient digestibility and plasma parameters of laying hens. *Animal Feed Science and Technology* 225, 20–26.

Liu, L., Li, Q., Yang, Y. and Guo, A. (2021) Biological function of short-chain fatty acids and its regulation on intestinal health of poultry. *Frontiers in Veterinary Science* 18, 736739.
Lu, J., Idris, U., Harmon, B., Hofacre, C., Maurer, J.J. and Lee, M.D. (2003) Diversity and succession of the intestinal bacterial community of the maturing broiler chicken. *Applied Environmental Microbiology* 69, 6816–6824.
Mäkeläinen, H., Forssten, S., Saarinen, M., Stowell, J., Rautonen, N. and Ouwehand, A.C. (2010) Xylo-oligosaccharides enhance the growth of bifidobacteria and Bifidobacterium lactis in a simulated colon model. *Beneficial Microbes* 1, 81–91.
Mahmood, T. and Guo, Y. (2019) Dietary fiber and chicken microbiome interaction: Where will it lead to? *Animal Nutrition* 6, 1–8.
Morgan, N.K., Wallace, A., Bedford, M.R. and Choct, M. (2017) Efficiency of xylanases from families 10 and 11 in production of xylo-oligosaccharides from wheat arabinoxylans. *Carbohydrate Polymers* 167, 290–296.
Morgan, N.K., Wallace, A., Keerqin, C., Wu, S.B. and Choct, M. (2018) Effect of arabinoxylans and xylo-oligosaccharides on net energy and nutrient utilization in broilers. *Animal Nutrition* 5, 56–62.
Morgan, N.K., Wallace, A., Bedford, M.R., Hawking, K.L., Rodrigues, I. *et al.* (2020) *In vitro* versus *in situ* evaluation of xylan hydrolysis into xylooligosaccharides in broiler chicken gastrointestinal tract. *Carbohydrate Polymers* 230, 115645.
Morgan, N.K., Gomes, G.A. and Kim, J.C. (2021) Comparing the efficacy of stimbiotic and a combination of xylanase and beta-glucanase, in broilers fed wheat-barley based diets with high or low AME. *Poultry Science* 100, 101383.
Morgan, N.K., Wallace, A. and Bedford, M.R. (2022) Improving sorghum digestion in broilers by targeting fermentation of xylan. *Animal Nutrition* 10, 198–206.
Mortimer, J.C., Miles, G.P., Brown, D.M., Zhang, Z., Segura, M.P. *et al.* (2010) Absence of branches from xylan in Arabidopsis gux mutants reveals potential for simplification of lignocellulosic biomass. *PNAS* 40, 17409–17414.
Moura, P., Barata, R., Carvalheiro, F., Gírio, F., Loureirodias M.C. and Esteves, M.P. (2007) In vitro fermentation of xylooligosaccharides from corn cobs autohydrolysis by Bifidobacterium and Lactobacillus strains. *LWT-Food Science Technology* 40, 963–972.
Niewold, T.A., Schroyen, M., Geens, M.M., Verhelst, R.S.B. and Courtin, C.M. (2012) Dietary inclusion of arabinoxylan oligosaccharides (AXOS) down regulates mucosal responses to a bacterial challenge in a piglet model. *Journal of Functional Foods* 4, 626–635.
Olukosi, O.A. and Bedford, M.R. (2019) Comparative effects of wheat varieties and xylanase supplementation on growth performance, nutrient utilization, net energy, and whole-body energy and nutrient partitioning in broilers at different ages. *Poultry Science* 98, 2179–2188.
Peng, F., Ren, J.L., Xu, F., Bian, J., Peng, P. and Sun, R.C. (2009) Comparative study of hemicelluloses obtained by graded ethanol precipitation from sugarcane bagasse. *Journal of Agricultural Food Chemistry* 57, 6305–6317.
Pourabedin, M., Guan, L. and Zhao, X. (2015) Xylo-oligosaccharides and virginiamycin differentially modulate gut microbial composition in chickens. *Microbiome* 3, 15.
Roberfroid, M.B. (2000) Prebiotics and probiotics: are they functional foods? *American Journal of Clinical Nutrition* 71, 1682S–7S.
Rycroft, C.E., Jones, M.R., Gibson, G.R. and Rastall, R.A. (2001) A comparative in vitro evaluation of the fermentation properties of prebiotic oligosaccharides. *Journal of Applied Microbiology* 91, 878–887.
Samanta, A.K., Kolte, A.P., Elangovan, A.V. Dhali, A., Senani, S. *et al.* (2017) Effects of corn husks derived xylooligosaccharides on performance of broiler chicken. *Indian Journal of Animal Science* 87, 640–643.
Scott, K.P., Martin, J.C., Duncan, S.H. and Flint, H.J. (2014) Prebiotic stimulation of human colonic butyrate-producing bacteria and bifidobacteria, *in vitro*. *FEMS Microbiology Ecology* 87, 30–40.

Servin, A.L. (2004) Antagonistic activities of lactobacilli and *bifidobacteria* against microbial pathogens. *FEMS Microbiology Reviews* 28, 405–440.
Sikandar, A., Zaneb, H., Younus, M., Khattak, F., Masood S. *et al.* (2017) Effect of sodium butyrate on performance, immune status, microarchitecture of small intestinal mucosa and lymphoid organs in broiler chickens. *Asian-Australian Journal of Animal Science* 30, 690–699.
Singh, A.K., Mishra, B., Bedford, M.R. and Jha, R. (2021) Effects of supplemental xylanase and xylooligosaccharides on production performance and gut health variables of broiler chickens. *Journal of Animal Science and Biotechnology* 12, 98.
Sun, Z.P., Lv, W.P., Yu, R.K., Li, J., Liu, H.H. *et al.* (2013) Effect of a straw-derived xylooligosaccharide on broilergrowth performance, endocrine metabolism, and immune response. *Canadian Journal of Veterinary Research* 77, 105–109.
Suo, H., Lu, L., Xu, G.-H., Xiao, L., Chen, X.-G. *et al.* (2015) Effectiveness of dietary xylo-oligosaccharides for broilers fed a conventional corn-soybean meal diet. *Journal of Integrative Agriculture* 14 (10), 2050–2057.
Swiatkiewicz, S., Koreleski, J. and Arczewska-Włosek, A. (2010) Effect of organic acids and prebiotics on bone quality in laying hens fed diets with two levels of calcium and phosphorus. *Acta Veterinaria Brno*, 79, 185–193.
Timbermont, L., Lanckriet, A., Dewulf, J., Nollet, N. Schwarzer, K. *et al.* (2010) Control of Clostridium perfringens-induced necrotic enteritis in broilers by target-released butyric acid, fatty acids and essential oils. *Avian Pathology* 39, 117–121.
Van Immerseel, F., Boyen, F., Gantois, I., Timbermont L., Bohez, L. *et al.* (2005) Supplementation of coated butyric acid in the feed reduces colonization and shedding of Salmonella in poultry. *Poultry Science* 84, 1851–1856.
Vazquez, M.J., Alonso, J.L., Dominguez, H. and Parajó, J.C. (2000) Xylo-oligosaccharides: manufacture and applications. *Trends in Food Science and Technology* 11, 387–393.
Wickramasuriya, S.S., Park, I., Lee, K., Lee, Y., Kim, W.H. *et al.* (2022) Role of physiology, immunity, microbiota, and infectious diseases in the gut health of poultry. *Vaccines* 10, 172.
Willemsen, L.E., Koetsier, M.A., Van Deventer, S.J.H. and Van Tol, E.A.F. (2003) Short chain fatty acids stimulate epithelial mucin 2 expression through differential effects on prostaglandin E(1) and E(2) production by intestinal myofibroblasts. *Gut* 52, 1442–1447.
Yuan, L., Li, W., Huo, Q., Du, C., Wang, Z. *et al.* (2018) Effects of xylo-oligosaccharide and flavomycin on the immune function of broiler chickens. *Peer J* 6, e4435.
Zeng, H., Xue, Y., Peng, T. and Shao, W. (2007) Properties of xylanolytic enzyme system in bifidobacteria and their effects on the utilization of xylooligosaccharides. *Food Chemistry* 101, 1172–1177.
Zhenping, S., Wenting, L., Ruikui, Y., Jia, L., Honghong, L. *et al.* (2013) Effect of a straw-derived xylooligosaccharide on broiler growth performance, endocrine metabolism, and immune response. *Canadian Journal of Veterinary Research* 77, 105–109.
Zhou, E. Pan, X.L. and Tian, X.Z. (2009) Application study of xylo-oligosaccharide in layer production. *Modern Applied Science* 3, 103–107.
Zhou, K., Wu, S., Guanghai, Q., Fu, Y., Weiwei, W. *et al.* (2021) Dietary supplemental xylooligosaccharide modulates nutrient digestibility, intestinal morphology, and gut microbiota in laying hens. *Animal Nutrition* 7, 152–162.
Zyla, K., Gogol, D., Koreleski, J., Swiatkiewicz, S. and Ledoux, D.R. (1999) Simultaneous application of phytase and xylanase to broiler feeds based on wheat: in vitro measurements of phosphorus and pentose release from wheats and wheat-based feeds. *Journal of the Science of Food and Agriculture* 79, 1832–1840.

CHAPTER 10

Paradigm Shifts in EU Animal Nutrition Legislation – Local and Global Implications

Elinor McCartney[1]*, Eliana Henríquez Rodriguez[1] and Dawn Scholey[2]

[1]*Pen & Tec Consulting S.L.U., Spain (now part of Argenta);* [2]*Poultry Research Unit, Nottingham Trent University, UK*

Abstract

The European Union (EU) has been driving changes in global food chain legislation over the last 50 years, notably since 2000, the so-called 'Brussels Effect'. The process started modestly in 1970, with the publication of an EU feed additive list, in which many entries were 'copy-pastes' from food additives. Then we had several food scandals in the 1990s, 'mad cows' (BSE, bovine spongiform encephalopathy) having the most impact, but also increasing concerns on antimicrobial resistance (AMR), soon to be associated with emerging 'superpathogens' such as *Escherichia coli* O157:H7 in cattle and *Salmonella* spp. in poultry.

Hence, the first paradigm shift came around 2000. The EU re-organized food chain legislation and shifted legal responsibility to DG Sante (Directorate General for Health and Consumers), thus promoting consumers, their concerns and preferences to the forefront of food chain legislation and creating an independent scientific body, the European Food Safety Authority (EFSA).

The EFSA has transformed scientific and regulatory approaches to the evaluation of quality, safety and efficacy of food-chain products, including food and feed additives, genetically modified organisms (GMOs) or genetically modified microorganisms (GMMs), applying exacting and ever-increasing standards of evaluation to the foods that we eat. EFSA's approaches have global implications, especially in non-EU countries with more modest regulatory structures that tend to use EFSA opinions and EU approvals as sound guarantees of product 'fit for purpose'.

*Corresponding author: elinor@pentec-consulting.eu

 Pre and Probiotics for Poultry Gut Health (eds H. Masey O'Neill, E. Burton and D. Scholey)
DOI: 10.1079/9781800622746.0010

The second paradigm shift is now upon us, with the advent of the EU transparency regulation, applied from March 2021. For various reasons, the transparency regulation will have local (European) and global effects on bird health, welfare and performance, nutritional and related components of poultry diets and will entail increased regulatory challenges for both innovative and existing products.

10.1 The Evolution of the EU – from the 1950s to the 2020s

The EU evolved in the decades following the aftermath of the Second World War, as European leaders sought to rebuild and maintain peace in Europe, initially through military and industrial collaborations, with developing and deeper future objectives in the spheres of financial, legal and political harmonization. Thus, in April 1949 the North Atlantic Treaty Organization (NATO) was created, originally an alliance of the USA, Canada and ten Western European countries. By 2020, NATO had 30 members, including 21 EU countries. In July 2022, NATO invited Finland and Sweden to join the organization and began the ratification process for these two countries. As of June 2022, three additional states have formally informed NATO of their membership aspirations: Bosnia and Herzegovina, Georgia and Ukraine. Kosovo also aspires to join NATO.

Regarding non-military developments, the Council of Europe was established in May 1949, to promote democracy, the protection of human rights and the rule of law. In 1951, six European countries formed the European Coal and Steel Community. These countries – Germany, France, Italy, the Netherlands, Belgium and Luxembourg – formed the core of what later became the EU. By 1958 the EEC (European Economic Community) came into being. The EEC, known as the 'Common Market', planned to work towards integration and economic growth through trade, with the founding objectives of free movement of goods, people, services and capital. Through many further developments, treaties and the creation of key European institutions, the EEC became what we know today as the EU. Table 10.1 depicts the Member States of the EU, currently 27 in number and with citizens numbering around 450 million. The UK, population 67 million, joined the EU in 1973 and left in January 2020 (EU Commission, 2022a), leaving behind many 'British' contributions to the EU, including English as an official EU and EFSA language.

10.2 EU Global Powers

In 2017, the world population was estimated by the United Nations at 7.6 billion (United Nations, 2017), hence the EU-27, with a population of only 450 million, represents only 6% of global citizens, compared with China (19% of global population) and India (18% of global population). Nevertheless, in 2017, gross domestic product (GDP) of the EU-27 represented 16.0% of the

Table 10.1. Member states of the European Union in 2020.

Country	Population (M)	Country	Population (M)	Country	Population (M)
Austria	8.9	France	67.5	Malta	0.5
Belgium	11.5	Germany	83.2	Netherlands	17.4
Bulgaria	7.0	Greece	10.7	Poland	38.0
Croatia	4.1	Hungary	9.8	Portugal	10.3
Cyprus	0.9	Ireland	5.0	Romania	19.3
Czechia	18.7	Italy	59.6	Spain	47.3
Denmark	5.8	Latvia	1.9	Sweden	10.3
Estonia	1.3	Lithuania	2.8	Slovakia	5.5
Finland	5.5	Luxembourg	0.6	Slovenia	2.1
Total (excluding UK) = ca 450 M		M = million		(UK)	(67.0)

Sources: EU Commission, 2022a; Eurostat, 2022.

world's GDP. In comparison, China and the USA were the world's two largest economies, with GDPs of 16.4% and 16.3%, respectively. India was the fourth largest world economy, with 6.7% of global GDP. The USA in 2017 with 330 million inhabitants represented 4.31% of the global population. Hence, after the USA and China, the EU is the third most powerful economy globally, despite its relatively small population. The EU provides strong leadership in consumer and food chain safety and EU food chain legislation tends to set global trends and standards.

It is not just food-chain legislation where the EU exerts influence on global standards. Chemical legislation such as REACH (Restriction, Evaluation, Authorisation and Restriction of Chemicals) and another EU institution, ECHA (European Chemicals Agency) have driven global approaches to the control of chemicals and their effects on the environment.

EU membership is formed by democracies, and apart from the basic freedoms of movement – goods, people, services and capital – EU citizens enjoy longevity, a good quality of life and social legislation designed to support vulnerable members of the community. The key mission of the EU Commission is to make EU citizens, happy, healthy and confident. The EU is far from perfect, but it is still a significant global player, and a valuable market for trading partners. EU consumers are increasingly discerning in terms of food quality, and EU supermarkets compete to buy the best food quality globally, sometimes reaching beyond the demands of EU legislation and EFSA opinions. Hence 'quality' to significant sectors of EU consumers today may mean:

- antibiotic-free poultry – neither antibiotics nor coccidiostats;
- GMO-free diets;
- better bird welfare, lower stocking densities, more humane slaughter;
- free-range birds;
- no-kill eggs (male embryos detected and eggs eliminated early in incubation); and
- novel, nutritious and tasty alternatives to eggs and poultry meat.

10.3 EU and EFSA Effects 2000–2010

As previously mentioned, during the last two decades of the 20th century several food scandals were key drivers of a major restructuring of EU food-chain legislation, leading to the creation of EFSA. The 'mad cow' scandal was probably the main instigator of the changes we saw in the early 2000s, since European citizens were horrified, first at the images of dairy cows affected with BSE, then with revelations about feeding processed animal proteins to herbivores and finally with the harrowing cases of human Creutzfeldt–Jakob Disease (CJD). European political leaders were determined to improve food-chain safety and charged the EU Commission to work with Member States to restore consumer confidence in the quality and safety of the food we eat. As a result, the whole EU approach to legislation in animal nutrition and production was reorganized and placed within DG SANCO, known then as the Directorate General for Health and Consumers, thus placing consumer health and welfare at the heart of the matter. DG SANCO today is DG SANTE, the Directorate General for Health and Food Safety.

The EU published the General Food Law in 2002, which was a key step to the creation of EFSA in February 2002. There were many legislative acts arising from the General Food Law, finally affecting everything '*From Farm to Fork*' (e.g. pesticides, feed ingredients, feed additives, food ingredients, food additives and all GMOs), but the initial acts were targeted in the area of animal nutrition and production, with a new feed additive regulation published in 2003 (Regulation EC N° 1831/2003, EU Commission, 2003). This regulation introduced sweeping changes, such as:

- completion of the ban on antibiotic growth promoters as feed additives;
- a review of the status of coccidiostats and histomonostats as feed additives;
- amino acids, urea and silage agents newly defined as feed additives;
- natural and synthetic flavours to be considered individually as feed additives;
- all existing feed additives to be notified to Member States and/or the EU Commission;
- re-authorization dossiers for all existing feed additives, to be submitted by latest November 2010;
- thereafter renewal of all feed additive authorizations required every 10 years;
- holder-specific authorizations granted for zootechnical feed additives, coccidiostats and histomonostats;
- live microorganisms used as feed additives to be fully characterized (e.g. silage inoculants and 'probiotics');
- producer microorganisms to be fully characterized for relevant feed additives (e.g. vitamins, amino acids and enzymes); and
- post-market monitoring introduced, including obligatory HACCP-based safety procedures (Hazard Analysis and Critical Control Points).

There were around 12,000 feed additives notified to Member States after publication of the 2003 feed additive regulation, with many duplicates (e.g. multiple notifications of amino acids, trace minerals, vitamins, etc.). A significant majority of notified feed additives were flavours, including flavours of natural origin. Today there are around 1000 entries in the EU feed additive register, of which an estimated 600 are flavours and about 200 of these are natural or nature-identical flavours. Many of these flavours, particularly those of natural botanical origin, were already being developed to help replace antibiotic growth promoters, and were not alone as innovative feed additives.

10.4 The Brussels Effect and Antimicrobial Resistance as Key Drivers for Change and Innovation

The 'Brussels Effect', first coined in 2012 by Anu Bradford, a Harvard-educated law scholar, and developed further in a book published in 2019, eloquently summarizes the global influence of EU legislation and institutions (Bradford, 2019):

> The Brussels Effect challenges the prevalent view that the European Union (EU) is a declining world power. It argues that notwithstanding its many obvious challenges, the EU remains an influential superpower that shapes the world in its image through a phenomenon called the "Brussels Effect." The Brussels Effect refers to the EU's unilateral power to regulate global markets. Without the need to resort to international institutions or seek other nations' cooperation, the EU has the unique ability among nations today to promulgate regulations that shape the global business environment, elevating standards worldwide and leading to a notable Europeanization of many important aspects of global commerce. Different from many other forms of global influence, the Brussels Effect entails that the EU does not need to impose its standards coercively on anyone —market forces alone are often sufficient to convert the EU standard into the global standard as multinational companies voluntarily extend the EU rule to govern their global operations. In this way, the EU wields significant, unique, and highly penetrating power to unilaterally transform global markets, including through its ability to set the standards in diverse areas such as competition regulation, data protection, online hate speech, consumer health and safety, or environmental protection. (Bradford, 2019)

An excellent example of EU power and the 'Brussels Effect' is the completion of the prohibition of antibiotic growth promoters, described below.

The EU had serious concerns on AMR by the 1990s, concerns that had being growing in the UK and Europe since the publication in 1969 of the Swann report on the use of antibiotics in animal husbandry and veterinary medicine (Swann, 1969). Indeed, concerns on antibiotic resistance were raised much earlier, almost from the discovery of the first antibiotic. Dr Alexander Fleming, a Scottish scientist, is credited with discovering that first antibiotic, penicillin, in 1928. For this work he was awarded the Nobel Prize in 1945. In his Nobel Prize acceptance speech he presciently warned about the dangers of antibiotic misuse: 'It is not difficult to make microbes resistant to penicillin in the laboratory by exposing them to concentrations not sufficient to kill them, and the same thing has occasionally happened in the body'.

As mentioned, the EU has power to approve and prohibit by legal acts, in particular 'Regulations', which are applied uniformly across all Member States and published in all 24 EU official languages, including English. This is a significant power, illustrated by the complete prohibition of antibiotic growth promoters in animal nutrition, implemented by the feed additive regulation and applied from 1 January 2006. In the USA, such complete prohibitions have been difficult to achieve with legislation, due to complex differences between US State laws and powerful industry lobbies. Indeed, in the USA the power against antibiotic use in poultry nutrition has arisen from market forces – changes in consumer preferences and demands that feed back to poultry producers through giant fast-food chains. Hence, now in the USA, two decades later than the EU, we have witnessed market trends for ABFs (antibiotic-free birds) and NAE (no antibiotics ever) poultry production.

10.5 EU and EFSA Effects 2010–2022 – EFSA Guidance Documents

Most of the period from 2000 to 2010 was used to draft new food chain regulations, to create EFSA and to prepare for the review and reauthorization of notified feed additives. EFSA used the experience of the first dossiers through the new system to develop EFSA guidance documents, that, in general, have become more challenging for applicants. A notable exception is the Qualified Presumption of Safety (QPS) approach for microorganisms used in the food chain, including feed additives. This is a list of microorganisms, frequently updated by the EFSA Biohazard Panel (EFSA 2022). QPS-listing exempts certain microorganisms from safety testing in target animals, consumers and the environment, but does not exempt from essential quality and efficacy data, nor from an assessment of user/worker safety. Typical QPS microorganisms are the lactic acid bacteria and common yeasts. EFSA has, however, frequently updated guidance on the taxonomy and characterization of live microorganisms used as feed additives, hence applicants must supply whole genome sequence data and current methodologies and database searches to identify any genetic elements of concern (e.g. toxins and genes coding for AMR).

Since 2010, EFSA has updated most guidance documents, summarized in Table 10.2. For successful feed additive approvals in the EU, it is important to follow current EFSA guidance in force at the date of dossier submission.

A good example of a recent and additional EFSA requirement for all feed additives are the 2021 EFSA guidance documents on nanoparticles (i.e. particles less than 100 nm in size). New and innovative nanotechnologies offer potential benefits to consumers and animals. For example nanoparticle technology applied to food and feed flavourings and spices, including salt or sugar, may offer benefits such as improved palatability at lower concentrations in food or feed products, thus offering potential safety benefits (e.g. lower salt or sugar diets, but still tasty!; or better efficacy at a lower dose, hence reduced residues from botanical preparations), or efficacy benefits (e.g. improved health and nutrition benefits due to enhanced bioavailability), or simply improved economy

Table 10.2. Key EFSA guidance documents – relevant for feed additives.

Guidance on studies concerning the safety of use of the additive for users/ workers	EFSA 2012
Guidance on the assessment of the safety of feed additives for the consumer	EFSA 2017a
Guidance on the assessment of the safety of feed additives for the target species	EFSA 2017b
Guidance on the identity, characterisation and conditions of use of feed additives	EFSA 2017c
Guidance on the assessment of the efficacy of feed additives	EFSA 2018a
Guidance on the characterisation of microorganisms used as feed additives or as production organisms	EFSA 2018b
Guidance on the assessment of the safety of feed additives for the environment	EFSA 2019
Guidance on the renewal of the authorisation of feed additives	EFSA 2020
Administrative guidance for the preparation of applications on additives for use in animal nutrition	EFSA 2021a
Guidance on risk assessment of nanomaterials to be applied in the food and feed chain: human and animal health	EFSA 2021b
Guidance on technical requirements for regulated food and feed product applications to establish the presence of small particles including nanoparticles	EFSA 2021c

of use (e.g. lower dose to achieve minimum flavour effect). But novelty and innovation raise safety concerns with scientific evaluators and regulators, hence new EFSA guidance and increased challenges for feed additive dossiers. Luckily, for highly soluble substances (e.g. salt or sugar), compliance with EFSA nanoparticle guidance is possible with a simple solubility test. Otherwise applicants must apply new technologies involving, among other data, specialized transmission and scanning electron microscopy techniques.

10.6 Feed Enzymes in Poultry Nutrition

The European poultry industry led innovation in the 1980s and 1990s through research and development of feed enzymes, originally targeted to enable birds to digest a wider variety of complex cereals, thus contributing to improved feed efficiency, increased flexibility in least-cost feed formulations and enhanced zootechnical performance. The European innovations in feed enzymes were originally driven by the much wider diversity of feed ingredients grown locally and used in poultry feeds, e.g. wheat and barley. New feed enzymes enabled utilization of higher percentages of these cereals without provoking wet litter problems, and subsequent developments in enzyme technology also allowed use of more exotic ingredients such as rye, sunflower, rapeseed and oats, as well as facilitating novel strategies such as feeding relatively high amounts of whole grains. Feed enzyme technologies were further developed to allow the use of enzymes in any type of processed poultry feed, either due to high enzyme tolerance to feed processing temperatures, enabled by geneticall enhanced producer strains, or post-processing spray applications of liquid enzymes.

For many years and until recently, particularly in the USA and Latin America, maize/soybean diets were the poultry diets of choice. For example, US regulators were reluctant to accept data generated from birds fed other types of diets. However, today, exogenous feed enzymes are recognized globally as helping to ameliorate removal of antibiotics from poultry diets, by improving digestibility and contributing to gut health and a much wider range of heterogenous diets and feeding practices are recognized as valid alternatives to the previous 'gold standard' maize/soybean diets.

10.7 Probiotics in Poultry Production

Another great innovation in feed additives during the 1980s and 1990s, was the arrival of 'probiotics'. The word 'probiotic' is not found in the feed additive regulation, or indeed anywhere else in EU legislation. In fact, use of the word 'probiotic' in foods and food supplements in the EU is controversial, since the EU consider that this implies an unauthorized nutritional and health claim. In another peculiar illogicality of EU food chain legislation, many species of live microorganisms may be added to foods and food supplements, without any requirement for a pre-marketing dossier, for example lactic acid bacteria used in dairy products, but all live microorganisms intended for use in animal feeds or drinking water must be authorized by the EU, but only after evaluation of strain safety, quality and efficacy by EFSA. The 'probiotic' vanguard in animal feeds was led by spore-forming *Bacillus* species, due to their resistance to heat, pressure and moisture during feed processing. Today there is a huge variety of probiotics used in animal feeds, including lactic acid bacteria, yeasts and other more exotic species, such as *Clostridium butyricum*. The new feed additive regulation helped the development of these innovations by maintaining their zootechnical status as feed additives, with enhanced protection of intellectual property through holder-specific authorizations. Thus, EU Member States do not raise objections to the marketing of live microorganisms as 'probiotics', provided they are duly authorized under the zootechnical category.

10.8 What's In A Name?

By the 1990s, the vocabulary used for functional groups within the zootechnical category of feed additives was already moving away from 'growth promotion' towards enhancement of zootechnical performance as 'gut flora stabilizers' or 'digestibility enhancers'. In general, 'probiotics' tend to be classed as 'gut flora stabilizers', whereas feed enzymes tend to be classed as 'digestibility enhancers'. However, in another European peculiarity, EFSA does not accept data showing enhanced beneficial bacteria as supportive of efficacy in the functional group 'gut flora stabilizer', but will accept data showing improved zootechnical performance (e.g. body weight, feed gain, feed efficiency) and/or suppression of undesirable microorganisms or enteropathogens (e.g. pathogenic *E. coli*, *Salmonella* spp., *Campylobacter* spp. and *Clostridium* spp.). Since 2003, the feed additive regulation has been amended to admit additional functional groups. In addition to the primary reasons of assuring the safety and quality of feed additives, the EU seeks to encourage innovation in areas of interest, such as feed additives that offer environmental and animal welfare benefits (Table 10.3). Innovative feed additives that target animal welfare benefits are classed as 'physiological condition enhancers'.

Legally, technological feed additives must act only in the feed and not in the animal. Scientifically, many preservatives, antioxidants, acidity regulators and hygiene improvers are known to act in the gut of the animal, exerting favourable effects. Examples are organic acids or lactic acid bacteria, which in any case require aqueous milieu to exert their acidifying action, since acidification is limited in low-moisture feeds. Mycotoxin binders in particular are known to act in the intestinal tract, after ingestion by the animal. Natural or

Table 10.3. EU Feed additive categories and functional groups, 2022.

Categories				
1. Technological	2. Sensory	3. Nutritional	4. Zootechnical	5. Coccidiostats, histomonostats
Functional groups				
Preservatives	Feed Colours	Vitamins Pro-vitamins	Digestibility enhancers	Coccidiostats, histomonostats
Antioxidants	Colours of food produce	Trace elements	Gut flora stabilizers	
Emulsifiers	Colours for ornamental fish and birds	Amino acids	Environmental improvers	
Stabilizers	Flavourings	Urea	Physiological condition stabilizers[b]	
Thickeners Gelling agents Binders Radionuclide controllers Anticaking agents Acidity regulators Silage additives Denaturants Mycotoxin reducers Hygiene improvers Other[a]			Other[a]	

[a]The applicant may propose a category.
[b]Feed additives that favourably affect the physiological condition of healthy animals, including their resilience to stress factors.

nature-identical flavours may also exert complex actions in the gut and after absorption to the bloodstream. The EU Commission and Member States resist attempts to exploit these features by claiming zootechnical benefits where no zootechnical authorization is in place. Measures taken include EFSA requirements for both *in vitro* and *in vivo* efficacy data for mycotoxin binders, setting upper limits on flavours to avoid zootechnical effects and claims, reallocation of some botanical active substances to zootechnical categories and general policing by Member States to avoid unlawful claims.

10.9 Prebiotics and Postbiotics

In contrast to probiotics (live microorganisms), prebiotics are typically substances such as oligosaccharides that are not digested by endogenous enzymes and may be fermented by beneficial intestinal microbes in the lower gut. Prebiotics may exert certain health benefits, such as improved gut motility (mild laxative effect) and better defence against undesirable microorganisms. Many such oligosaccharides are considered as feed materials, not feed additives, thus no pre-market dossier is required. Limited nutritional and physiological claims are permitted under the EU feed regulation. The main problem is that, in the absence of an independent EFSA evaluation of quality, safety and efficacy, market competition may drive down prices and compromise product quality. Under an EFSA assessment, feed additives must demonstrate adequate efficacy at a minimum recommended dose and over a minimum duration of feeding. This is one reason why a positive EFSA opinion is highly valued in the EU and globally.

Postbiotics have been defined recently by the International Scientific Association for Probiotics and Prebiotics (ISAPP) as 'preparations of inanimate microorganisms and/or their components that confer a health benefit on the host' (Salminen *et al.*, 2021). The ISAPP considered that effective postbiotics must contain inactivated microbial cells or cell components, with or without metabolites, that contribute to observed health benefits. As is the case with prebiotics, postbiotics are classed mainly as feed materials, hence there is no independent EFSA evaluation, especially important for quality and efficacy.

10.10 Synbiotics

Synbiotics are defined by the ISAPP as 'a mixture comprising live microorganisms and substrate(s) selectively utilized by host microorganisms that confers a health benefit on the host' (Swanson *et al.*, 2020). Synbiotics can also fit under the feed additive regulation since feed additives are defined as 'substances, microorganisms or preparations added to feed or water'. Feed additives may be presented as complex mixtures – 'preparations' – that contain several active substances or active agents, including live microorganisms, diluents, excipients, carriers and coating agents.

10.11 Bacteriophages

Bacteriophages (viral phages that infect and destroy bacteria) are more recent types of feed additives, emerging from well-established innovative technologies used in Eastern Europe in both humans and animals. A pioneer bacteriophage dossier has been evaluated by EFSA and succeeded in terms of safety and quality, most likely pending supplementary efficacy data (EFSA, 2021d). Such is often the case with pioneer feed additives – it takes a few attempts to generate the right efficacy data to satisfy EFSA.

10.12 Feed, Feed Additives and PARNUTS Claims

The EU feed regulation permits certain claims on feeds, essentially nutritional and physiological in nature, provided they are non-veterinary and supported by data: 'The labelling and the presentation of feed materials and compound feed may draw particular attention to the presence or the absence of a substance in the feed, to a specific nutritional characteristic or process or to a specific function ....'.

Feeds that contain feed additives may also make additional claims associated with the authorized feed additive category, for example improved zootechnical performance. EU Member States are responsible for policing claims made on feeds and may vary in the stringency of the control and associated penalties for transgression. Feed business operators or Member States may consult with or appeal to the EU Commission in 'grey areas', usually involving borderline veterinary claims in relation to treating, preventing or curing disease.

The PARNUTS regulation covers feeds for particular nutritional purposes (i.e. PARticular NUTritional purposeS) and allows certain claims on feeds that meet specified characteristics. Interestingly, PARNUTS dossiers are initially reviewed by the EU Commission and Member States and only referred to EFSA where safety questions arise.

Many PARNUTS claims are related to disease conditions: for example, reduction of the risk of milk fever (hypocalcaemia) in dairy cows. Most PARNUTS claims are for companion animals, driven by the diverse ranges of high-quality pet foods marketed, targeted at ameliorating or reducing the risk of various disease conditions (e.g. chronic kidney disease, kidney stones, various digestive disorders, skin dermatosis, obesity). There are few PARNUTS claims for poultry, possibly due to the highly integrated nature of poultry production and the economic pressure to provide cost-efficient diets (Table 10.4). Nevertheless, there is scope for creative innovation in the form of complementary feeds presented as coloured gels for spray application to hatching or newborn chicks at the hatchery or on arrival to farms. These PARNUTS feeds may be formulated to contain 'probiotic' gut flora stabilizers and other essential nutrients, as well as water. Such feeds have been proven to be highly effective in reducing early post-hatch chick mortality, probably by achieving early hydration of chicks and helping to support gut flora, health and nutritional needs in the first few days of life.

Table 10.4. PARNUTS feeds for avian and poultry species.

PARNUT	Essential nutrients	Species	Labelling	Duration	Other
50. Support preparation for reproduction	High vitamin A and/or vitamin D or high selenium and/or zinc and/or minimum vitamin E	Birds (i.e. all avian species)	Names and concentrations of vitamins and trace elements	During oestrus (females) and reproductive activity (males)	The complementary feed may contain selenium, zinc, vitamin A and D in a concentration higher than 100 times the relevant fixed maximum content in complete feed
52. Support for nutritional imbalances in dietary transition	Minimum supply of selenium, and/or zinc, and/or copper and/or vitamins A, and/or D and/or E	Poultry (i.e. poultry intended for meat or egg production)	Names and concentrations of vitamins and trace elements	2 to 15 days	The complementary feed may contain selenium, zinc, copper, vitamin A and vitamin D in a concentration higher than 100 times the relevant fixed maximum content in complete feed, but without passing maximum limits in final complete feed.
54. Support the regeneration of the skin and appendages	Minimum supply of zinc, and high copper and/or iodine and/or selenium and/or vitamins B6, and/or A, and/or E, and/or methionine and/or cystine	Poultry	Names and concentrations of vitamins and trace elements	Up to 8 weeks	The complementary feed may contain zinc, copper, selenium, iodine and vitamin A in a concentration higher than 100 times the relevant fixed maximum content in complete feed, but without passing maximum limits in final complete feed.
65. Compensation for malabsorption	Low saturated fatty acids and high fat-soluble vitamins	Poultry excluding geese and pigeons	% saturated fatty acids: total fatty acids Total vitamins A, D, E and K	1st 2 weeks post-hatching	

66. Reduction of the risk of fatty liver syndrome	Low energy and high proportion of metabolizable energy from lipids with high level of polyunsaturated fatty acids	Laying hens	EC calculated energy % metabolizable energy from lipids Content of polyunsaturated fatty acids	Up to 12 weeks	
72. Stabilization of the physiological digestion	Feed additives approved as gut flora stabilizers	Avians for which gut flora stabilizers are approved	Name and added amount of gut flora stabilizer	Up to 4 weeks	Label must state: 'In case of risk of, during periods of, or recovery from digestive disturbances' Complete feed must respect maximum legal inclusion rates of gut flora stabilizer

10.13 REFIT and the Transparency Regulation

The latest paradigm shift in EU legislation in the area of animal nutrition involves the revision of the feed additive regulation as part of a continual REFIT programme (Regulatory Fitness and Performance Programme) and the transparency regulation.

Regarding REFIT, the EU Commission update intends to make it easier to bring sustainable and innovative additives to market and streamline the authorization process without compromising health and food safety. REFIT is intended to make livestock farming more sustainable and reduce its environmental impact in line with the EU farm to fork strategy. The REFIT process was open to public consultation of all stakeholders and the current schedule envisages adoption of the EU Commission legislative proposal in late 2022. From the perspective of feed additive registrations, there is generally universal support for basic concepts such as:

- The EU feed additive registration process plays an important role in assuring safety, quality and efficacy of feed additives.
- Support of reduction of *in vivo* safety testing on live animals and use of *in vitro* alternatives where possible.

Nevertheless, typical major concerns and suggestions of manufacturers and applicants of feed additive dossiers include:

- There must be a balance between data requirements, protection of investment and intellectual property, as well as other considerations, such as animal and environmental welfare.
- Holder-specific authorizations for all feed additives will not only protect the intellectual property (IP) of applicants, but also strengthen post-market monitoring, and so enhance food chain safety.
- Holder-specific authorizations encourages innovation from all players, including small- and medium-size enterprises, who will be motivated to invest in the registration of feed additives, and enjoy good protection of intellectual property as a reward for their efforts and investment.
- There is room for more pragmatic approaches to certain safety testing, e.g. proteinaceous substances such as enzymes and live microorganisms are likely to be sensitizing in some individuals and, where there are practical difficulties to test *in vitro* (e.g. due to lack of solubility), then the substance can be listed on the safety data sheets as 'potential ssensitizer', without further *in vivo* testing.
- Regarding efficacy, there is scope for reduction of efficacy testing across all feed additives, to a single conclusive efficacy test (or metanalysis of two or more pooled efficacy studies).
- There is further scope to extend extrapolation of efficacy across animal categories and species.
- Shorter label texts are suggested, with detailed safety information supplied in the safety data sheets.
- Existing REACH/CLP legislation (Registration, Evaluation, Authorisation and Restriction of Chemicals/Classification, Labelling and Packaging) can

be used to ensure user/worker safety for relevant feed additives and chemicals, thus avoiding duplication of different areas of legislation.

The transparency regulation (Regulation [EU] 2019/1381; EU Commission, 2019) is probably the major paradigm shift in EU food-chain legislation since the creation of EFSA and the general food law regulation of 2002, now updated (EU Commission, 2022b). This new transparency regulation arose from a successful European Citizen's Initiative in relation to safety and other concerns regarding pesticides, particularly glyphosate, and was applied from 27 March 2021.

The transparency regulation aims to improve the transparency of EU food chain risk assessments, as well as the reliability, objectivity and independence of the studies used by EFSA. The powers and financing of EFSA are increased significantly and all EU Member States are represented on the EFSA Management Board from July 2022. The transparency regulation affects all 'REPRO' (regulated products), including:

- plant protection products;
- feed additives;
- novel foods;
- nutrition and health claims on foods;
- food contact materials;
- food additives, enzymes and flavourings, including smoke flavourings; and
- genetically modified organisms (cultivation and food/feed uses).

The transparency regulation mandates significant changes to EFSA's evaluation and EU processing of submitted REPRO dossiers:

- There is public access to scientific studies and information submitted to EFSA by applicants.
- EFSA conducts public consultations of stakeholders to identify any addition relevant data.
- All submitted files to be in e-format, word-searchable, capable of downloading and printing.
- EFSA may commission additional, independent confirmatory studies.
- Obligatory pre-notification of all studies planned and conducted by applicants, and by EU-based laboratory/research facilities, with penalties for late notifications or failures to notify. Study notifications must be lodged on the EFSA Connect Portal, a new platform (https://connect.efsa.europa.eu/RM/s/).
- Confidential information is limited to justified cases and decided by EFSA, for example:
    - manufacturing data, unless relevant for safety;
    - in-house developed analytical methods for feed additive impurities unless relevant for safety;
    - feed additive composition (excluding active substance/s), unless relevant for safety;
    - efficacy study plans;
    - General Data Protection Regulation (GDPR) personal data;
    - personnel and facilities involved in laboratory animal safety testing;

   - commercial links between applicants and commercial partners (e.g. producers/importers); and
   - market shares and business strategy of applicants.
- Feed additive dossiers must now be submitted electronically via the ESFCP: (E-Submission Food Chain Platform).
- The default setting is that submitted files are not confidential, hence most files must be additionally processed to claim and justify confidential parts, as well as earmarking then redacting GDPR and other confidential parts of files.

In addition to the considerable administrative complexity and challenges imposed by the transparency regulation, EFSA now apply current guidance documents rigorously, notably on target animal safety testing and efficacy studies. EFSA's current approaches are based on the determination to apply 'regulatory science', i.e. current EFSA standards, to submitted dossiers, as well as a rigid application of the principle that feed additives must show efficacy in animals in good health. Some recent examples are:

- EFSA now require that animal studies are carried out strictly under EU conditions of health, welfare and hygiene and may reject studies that do not comply fully with EU legislation on heating, lighting, stocking density and litter conditions. Thus, studies carried out in the USA or other non-EU countries may be rejected for non-compliance.
- With respect to challenge studies, these have previously been accepted by EFSA and in theory will still be accepted by EFSA. It has been a frequent practice in broiler efficacy studies to introduce minor stresses, such as used litter or heat stress, but EFSA may reject these studies unless the birds start under clean litter and other high welfare conditions, prior to applying the stress factors. It is also advisable to clarify in the study design that this is a challenge study, in order to show benefits of the feed additive in birds that may suffer certain common stressors during rearing, e.g. subclinical coccidial challenge.

10.14 Conclusion

The initial years of the 21st century have brought sweeping changes to EU legislation affecting the food chain, with both local and global implications. The advent of the transparency regulation will ensure that the EU and EFSA continue to mark the highest standards of feed additive evaluation, requiring a high degree of expertise and, above all, high-quality feed additives, to achieve successful authorization in Europe.

References

Bradford, A. (2019) *The Brussels Effect: How the European Union Rules the World* (New York, 2020; online edn, Oxford Academic, 19 December 2019). Available at: https://doi.org/10.1093/oso/9780190088583.001.0001 (accessed 2 August 2022).

EFSA (2012) EFSA Panel on Additives and Products or Substances used in Animal Feed (FEEDAP), Guidance on studies concerning the safety of use of the additive for users/workers. *EFSA Journal* 10, 2539. doi: 10.2903/j.efsa.2012.2539.

EFSA (2017a) EFSA FEEDAP Panel (EFSA Panel on Products or Substances used in Animal Feed), Guidance on the assessment of the safety of feed additives for the consumer. *EFSA Journal* 15, 5022. doi: 10.2903/j.efsa.2017.5022.

EFSA (2017b) EFSA FEEDAP Panel (EFSA Panel on additives and products or substances used in animal feed), Guidance on the assessment of the safety of feed additives for the target species. *EFSA Journal* 15, 5021. doi: 10.2903/j.efsa.2017.5021.

EFSA (2017c) EFSA FEEDAP Panel (EFSA Panel on additives and products or substances used in animal feed), Guidance on the identity, characterisation and conditions of use of feed additives. *EFSA Journal* 15, 5023. doi: 10.2903/j.efsa.2017.5023.

EFSA (2018a) EFSA FEEDAP Panel (EFSA Panel on additives and products or substances used in animal feed), Guidance on the assessment of the efficacy of feed additives. *EFSA Journal* 16, 5274. doi: 10.2903/j.efsa.2018.5274.

EFSA (2018b) EFSA FEEDAP Panel (EFSA Panel on Additives and Products or Substances used in Animal Feed), Guidance on the characterisation of microorganisms used as feed additives or as production organisms. *EFSA Journal* 16, 5206. doi: 10.2903/j.efsa.2018.5206.

EFSA (2019) EFSA FEEDAP Panel (EFSA Panel on Additives and Products or Substances used in Animal Feed), Guidance on the assessment of feed additives for the environment. *EFSA Journal* 17, 5648. doi: 10.2903/j.efsa.2019.5648.

EFSA (2020) EFSA FEEDAP Panel (EFSA Panel on Additives and Products or Substances used in Animal Feed), Guidance on the renewal of the authorisation of feed additives. *EFSA Journal* 19, 6340. doi: 10.2903/j.efsa.2021.6340.

EFSA (2021a) European Food Safety Authority, Administrative guidance for the preparation of applications on additives for use in animal nutrition. *EFSA supporting publication* EN-6508. doi: 10.2903/sp.efsa.2021.EN-6508.

EFSA (2021b) EFSA Scientific Committee, Guidance on risk assessment of nanomaterials to be applied in the food and feed chain: human and animal health. *EFSA Journal* 19, 6768. doi: 10.2903/j.efsa.2021.6768.

EFSA (2021c) EFSA Scientific Committee, Guidance on technical requirements for regulated food and feed product applications to establish the presence of small particles including nanoparticles. *EFSA Journal* 19, 6769. doi: 10.2903/j.efsa.2021.6769.

EFSA (2021d) EFSA FEEDAP Panel (EFSA Panel on Additives and Products or Substances used in Animal Feed), Scientific Opinion on the safety and efficacy of a feed additive consisting on the bacteriophages PCM F/00069, PCM F/00070, PCM F/00071 and PCM F/00097 infecting Salmonella Gallinarum B/00111 (Bafasal®) for all avian species (Proteon Pharmaceuticals S.A.). *EFSA Journal* 19, 6534. doi: 10.2903/j.efsa.2021.6534.

EFSA (2022) BIOHAZ Panel (EFSA Panel on Biological Hazards), Statement on the update of the list of QPS-recommended microbiological agents intentionally added to food or feed as notified to EFSA 16: Suitability of taxonomic units notified to EFSA until March 2022. *EFSA Journal* 20, 7408. doi: 10.2903/j.efsa.2022.7408.

EU Commission (2003) Regulation (EC) N° 1831/2003 of the European Parliament and of the Council of 22 September 2003 on additives for use in animal nutrition. (OJ L 268, 18.10.2003, p. 29). Available at: http://data.europa.eu/eli/reg/2003/1831/2021-03-27 (accessed 14 April 2023).

EU Commission (2019) Regulation (EU) 2019/1381 of the European Parliament and of the Council of 20 June 2019 on the transparency and sustainability of the EU risk assessment in the food chain and amending Regulations (EC) N° 178/2002, (EC) N° 1829/2003, (EC) N° 1831/2003, (EC) N° 2065/2003, (EC) N° 1935/2004, (EC) N° 1331/2008, (EC) N° 1107/2009, (EU) 2015/2283 and Directive 2001/18/EC. Available at: http://data.europa.eu/eli/reg/2019/1381/oj (accessed 14 April 2023).

EU Commission (2022a) History of the EU. Available at: https://european-union.europa.eu/principles-countries-history/history-eu_en (accessed 24 July 2022).

EU Commission (2022b) Consolidated Regulation (EC) N° 178/2002 of the European Parliament and of the Council of 28 January 2002 laying down the general principles and requirements of food law, establishing the European Food Safety Authority and laying down procedures in matters of food safety (OJ L 31, 1.2.2002, p. 1). Available at: http://data.europa.eu/eli/reg/2002/178/2022-07-01 (accessed 14 April 2023).

Eurostat (2022) Eurostat. Available at: https://ec.europa.eu/eurostat/statistics-explained/index.php?title=Purchasing_power_parities_in_Europe_and_the_world#Shares_in_world_GDP (accessed 24 July 2022).

Salminen, S., Collado, M.C., Endo, A., Hill C., Lebeer, S. *et al.* (2021) The International Scientific Association of Probiotics and Prebiotics (ISAPP) consensus statement on the definition and scope of postbiotics. *Nature Reviews Gastroenterology and Hepatology* 18, 649–667. doi: 10.1038/s41575-021-00440-6.

Swann, M. (1969) The use of antibiotics in animal husbandry and veterinary medicine (Swann report). HMSO, London.

Swanson, K.S., Gibson, G.R., Hutkins, R. Reimer, R.A., Reid, G. *et al.* (2020) The International Scientific Association for Probiotics and Prebiotics (ISAPP) consensus statement on the definition and scope of synbiotics. *Nature Reviews Gastroenterology and Hepatolology* 17, 687–701. doi: 10.1038/s41575-020-0344-2.

United Nations (2017) World Population Prospects: The 2017 Revision. Available at: https://www.un.org/development/desa/pd/sites/www.un.org.development.desa.pd/files/files/documents/2020/Jan/un_2017_world_population_prospects-2017_revision_databooklet.pdf (accessed 4 April 2023).

CHAPTER 11
Potential Application of Bacteriophage Therapy to Improve Poultry Health

Anisha M. Thanki*, Steve P.T. Hooton and Martha R.J. Clokie

Department of Genetics and Genome Biology, University of Leicester, Leicester, UK

Abstract

Antibiotics are frequently used in poultry production, but broad-spectrum antibiotics are problematic in driving antimicrobial resistance (AMR) within the food chain. Consequently, alternative antimicrobials are needed to treat bacterial infections and bacteriophages (phages) could provide an effective approach. Phages are natural parasites of bacteria and have been used therapeutically to treat both human and animal infections, including within poultry for over a hundred years. This chapter will provide an overview on the potential applications of phages within the poultry industry to reduce pathogen levels throughout the production process. It will include discussion on what phages are, where they are found, experimental poultry phage efficacy studies, phage products currently being sold commercially, the impact of phages on the poultry microbiome and advantages and limitations of using phages therapeutically. The research discussed in this chapter demonstrates the safety and efficacy of phages to reduce pathogen abundance within poultry production and supports their use.

11.1 Introduction

Antimicrobial resistance (AMR) is a persistent global health issue and, as antibiotics become less efficient at treating bacterial infections, AMR infections continue to cause high levels of morbidity and mortality. In 2019, an estimated

*Corresponding author: at263@leicester.ac.uk

DOI: 10.1079/9781800622746.0011

4.95 million deaths were associated with bacterial AMR infections (Murray *et al.*, 2022) and by 2050 could lead to 10 million deaths worldwide if the problem is not curtailed (O'Neill, 2016). There are connections between the use of antimicrobials in animal production and the spread of AMR in clinical settings as animal production utilizes 70% of total antimicrobials produced. Thus, multiple countries including the USA and the European Union (EU) have banned the use of antibiotics as growth promoters; however, some countries continue to use antibiotics in animal production to improve performance. Furthermore, as animal production expands globally, antibiotic usage has been predicted to increase to 67% by 2030, particularly in South Africa, Brazil, Russia, China and India, which ultimately will lead to the further spread of AMR bacteria within the food chain (Hedman *et al.*, 2020).

Within animal production, high levels of AMR have been observed in bacterial species linked to human food poisoning and recently the European Food Safety Authority (EFSA) reported multi-drug resistant (MDR) *Salmonella* spp. were recovered from broiler, pig, calf and turkey carcasses at levels of 53.6%, 43.3%, 23.1% and 19.1%, respectively. The highest levels of MDR were found in broilers and, from 984 *Salmonella* strains used in the analysis, 51.2% of strains were resistant to three or more antimicrobials commonly used to treat the infection and 20.5% of strains were resistant to one or two antimicrobial classes. Similarly, 3382 *Campylobacter* strains were isolated from broilers in the same study, of which 76.2% of strains were resistant to one or more antimicrobials. Resistance levels were highest against clinically important antibiotics nalidixic acid, ciprofloxacin and tetracycline. Likewise, in 4709 *Escherichia coli* strains isolated from broilers, 50% were resistant to ampicillin and ciprofloxacin. These resistant strains are spreading though the human food chain and thus limiting the activity of clinically important antimicrobials (EFSA and ECDC, 2022).

Controlling the use of antibiotics in animal production to limit AMR without compromising animal health and performance is of utmost importance and alternative antimicrobials are much needed. Significant investments from government bodies and the private sector are being made in this growing area of research into innovative alternative treatments, which includes probiotics, prebiotics, organic acids, enzymes and phytogenics (phytobiotics) (Gadde *et al.*, 2017). Natural predators of bacteria called bacteriophages (phages for short) are a particularly promising alternative antimicrobial and this chapter will discuss potential application of phages within the poultry industry.

11.2 Overview on Phages

Phages are viruses that specifically infect and kill bacteria and, importantly in most cases, have no negative effect on animal or human cells (Żbikowska *et al.*, 2020). Phages were independently discovered over a century ago by Frederick Twort and Felix d'Herelle. Since their discovery, they have been widely used in the treatment of bacterial infections (referred to as phage therapy) in both animals and humans in Eastern European countries, where the leading phage treatment centres in the world are located: the Eliava Institute of Bacteriophages, Microbiology and Virology in Tbilisi, Georgia and the Bacteriophage Therapy Unit of the Ludwik Hirszfeld Institute of Immunology and Experimental Therapy in Wroclaw, Poland (Żaczek *et al.*, 2020; Chanishvili *et al.*, 2022). In comparison, other countries have relied on antibiotics but, owing to the prevalence in infections caused by MDR bacteria, phage research is growing worldwide.

Phages are categorized by their morphology, their genome sequence and genome composition (if the phages have a DNA or RNA genome). Phages are a thousand times smaller than bacterial cells and their morphologies can only be viewed using a transmission electron microscope (TEM) at ~150,000 times magnification. Historically, phages were grouped based on their morphology and from over 5000 phages that have been viewed under a TEM, 96% of phages had tails and were part of the order *Caudovirales*. The order was divided into *Siphoviridae* (61% of phages), *Myoviridae* (25%) and *Podoviridae* (14%) families and examples of each are shown in Fig. 11.1 (Ackermann, 2009). Over the past decade, the number of phages being isolated is increasing exponentially and subsequently phage taxonomy has been updated. Currently there is a new order called *Tubulavirales*, ten new families and twenty-two new subfamilies (Adriaenssens *et al.*, 2020).

11.3 Phage Infection Cycles

Phages are obligate parasites of bacteria and can infect via two main cycles: lytic or lysogenic. Phages first bind to a receptor(s) on the surface of the bacterial cells then inject their DNA and take over metabolism of the bacterial cells to replicate themselves if they are lytic, which leads to lysis of the cell and the release of newly assembled phages. The new progeny can go on to infect more of the target bacterial cells and the cycle is repeated. In contrast, in the

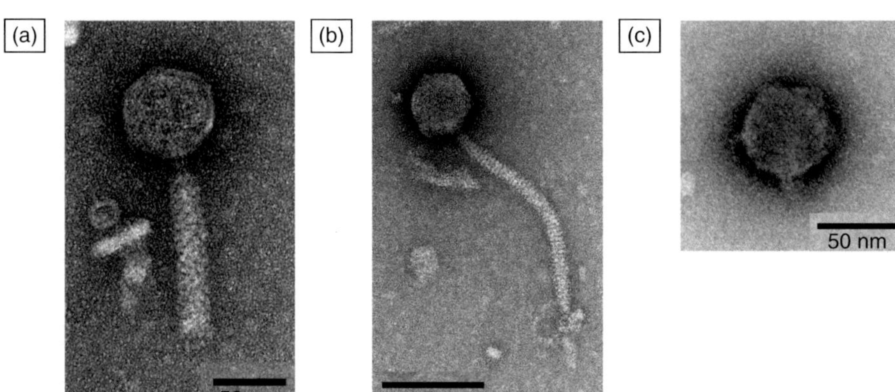

Fig. 11.1. TEM images showing typical morphology of phages from (a) *Siphoviridae*, (b) *Myoviridae* and (c) *Podoviridae* families.

lysogenic cycle, phage DNA becomes incorporated into the bacterial genomes and is then replicated along with the host, but under stressful conditions the phage can also enter the lytic cycle (Clokie *et al.*, 2011; Thanki *et al.*, 2021). Sequencing the phage genome identifies the life cycle the phage follows and as phages following the lytic cycle kill their target bacterial cells directly, they are the preferred phages for therapeutic applications (Shen and Millard, 2021).

11.4 Phage Isolation and Testing their Lytic Spectrum

Phages are globally ubiquitous and in total they outnumber bacteria cells by ten times. Humans and animals are continuously exposed to phages because they are found in food, seawater and mud and are part of our healthy gut microbiome. Theoretically, there is at least one type of phage that infects every strain of bacteria and it is a matter of isolating infecting phages. Phages can be isolated from environments in which their target bacteria naturally thrive. For example, the bacterial pathogens *Salmonella* and *Campylobacter* are commonly found on farms and phages against both have been isolated from chicken and pig faeces, sewage, slurry and broiler house samples (Jäckel *et al.*, 2019; Thanki *et al.*, 2019). Interestingly, the majority of phages isolated against both pathogens have been from samples collected from healthy animals, where phages are probably playing a natural role in microbial defence.

The protocol used to isolate phages from environmental samples (referred to as the sample) involves the following steps and the steps are shown in Fig. 11.2 (Wommack *et al.*, 2009):

1. The sample is mixed with a target bacterial strain or multiple strains in a liquid media, which allows growth of the target bacteria. Bacterial strains are added at this step to amplify any phages that are present and capable of lysing the target strains. If phages are present, they will probably be present at a low concentration and to increase their concentration they will use the added strains to replicate to the level they can be detected.

2. The mixture is incubated overnight in conditions that favour bacterial growth (appropriate temperature, aerobic/anaerobic atmosphere, presence or absence of divalent cations such as magnesium sulphate and calcium chloride that can help phages bind to their target strains) and consequently facilitate phage replication.

3. The mixture is spun down and the environmental sample/bacteria gather at the bottom of the tube as they are heavier in weight. The phages remain in the liquid, which is then filtered (e.g. through a 0.2 μm pore size filter) and, as phages are smaller in molecular size, they will pass through the filter whilst the bacterial cells will not.

4. The filtered sample is screened for phages by mixing this sample with the target bacterial strain, which is then added to semi-solid media and poured on to petri-dishes with a solid media and incubated overnight.

5. If phages are present at a high concentration in the sample, clearing will be observed, which occurs when the phage has completely lysed the strain and

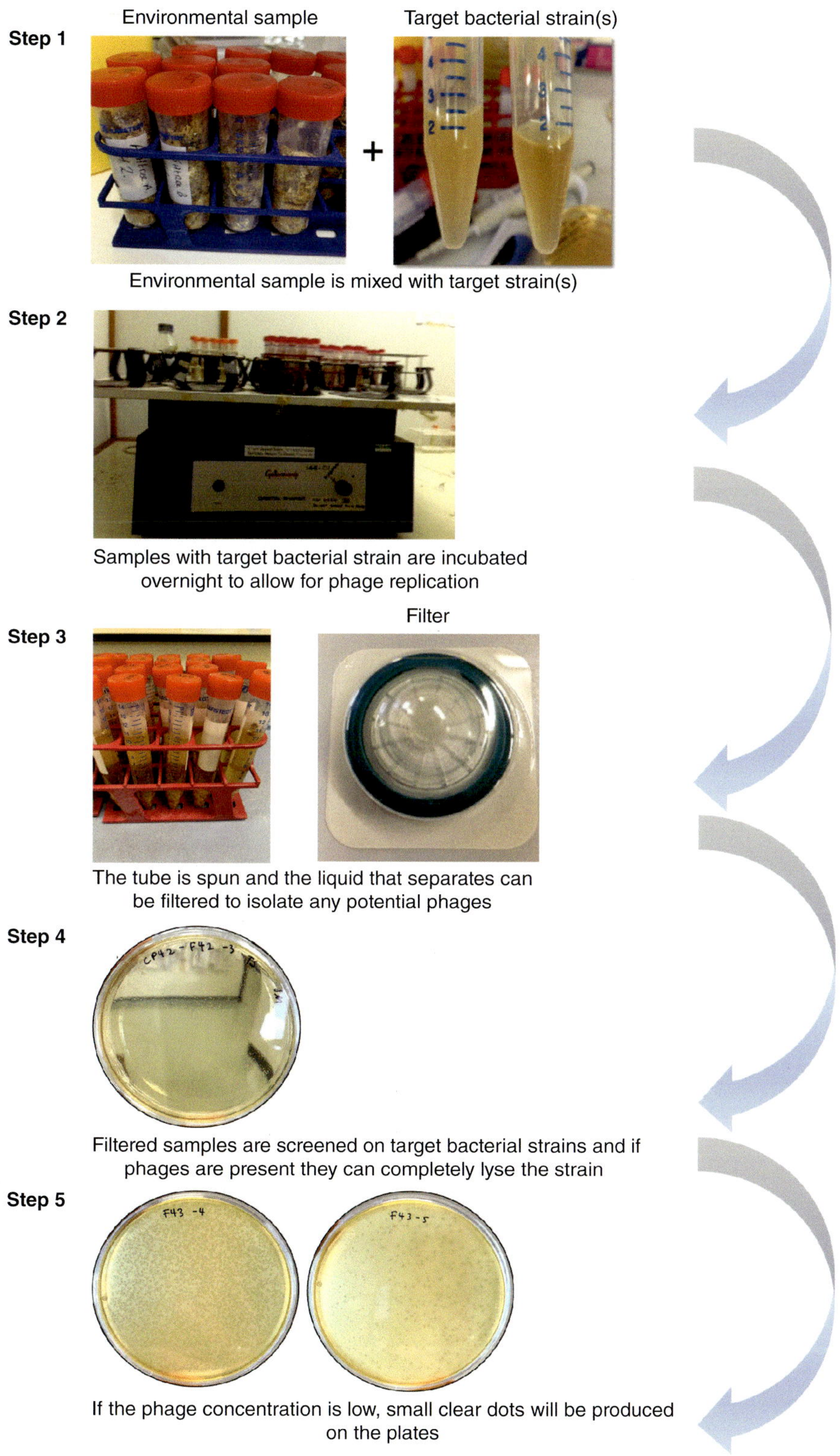

Fig. 11.2. The steps followed to isolate phages from environmental samples (such as faecal, river, sewage and slurry samples). A description of the steps involved is included in the text.

prevented it from growing. If phages are present at lower concentrations, plaques will be seen, which appear as clear dots on the plate.

After phages have been isolated, the next step is to determine their lytic spectrum. Typically, phages are bacterial species-specific and, for example, *Salmonella* phages are unlikely to be capable of infecting strains of *Clostridium perfringens*. Consequently, unlike antibiotics, phages can provide a targeted treatment against a single bacterial species without disrupting other species in the microbiome. Phages can infect multiple sub-groups of the same bacterial species and are referred to as broad host range phages. A collection of *Salmonella* phages isolated by our research group in 2017 have broad host range and can infect representative strains including MDR strains from multiple *Salmonella* serotypes associated with infections in pigs and chickens, including *S.* Typhimurium and its monophasic variants; *S.* Derby; *S.* Enteritidis and *S.* Infantis (Thanki *et al.*, 2019; Nale *et al.*, 2021). In comparison, the *Clostridium difficile* phage phiCDHM4, which was also isolated by our research group, has a narrow host range and can only infect four strains of thirty-two screened (Nale *et al.*, 2016). To overcome narrow specificity, phages can be combined as 'cocktails' to increase their lytic spectrum and such 'cocktails' have been designed against the common poultry pathogens *Salmonella* (Thanki *et al.*, 2022a,b), *Campylobacter*, *E. coli*, *Staphylococcus aureus* (Żbikowska *et al.*, 2020) and *C. perfringens* (Miller *et al.*, 2010).

11.5 Potential Application of Phages in the Poultry Industry

Phages can be used at multiple points through the farm to fork process from feed manufacturing, poultry farming, processing to production. Phages can be delivered pre-slaughter in drinking water or in feed and post-slaughter phages can be sprayed or dipped onto carcases or meat products. Figure 11.3 summarizes three key areas of phage application to reduce foodborne pathogens

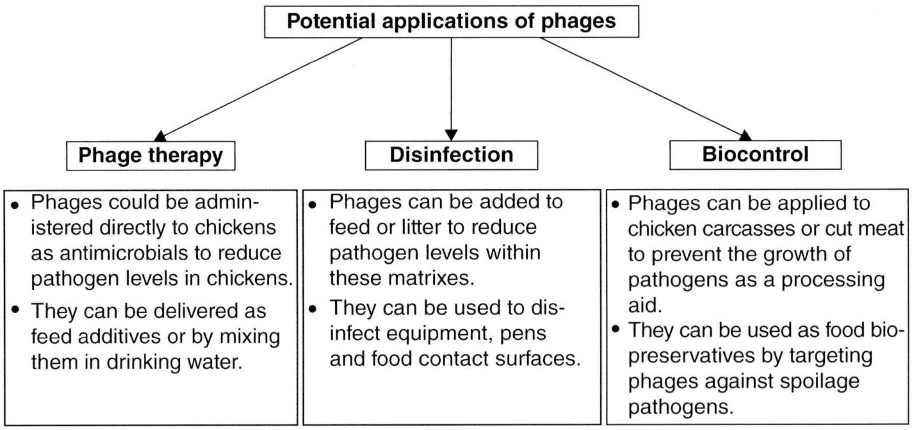

Fig. 11.3. Areas in which phages can be applied in the poultry industry.

within the food chain. The areas are: phage therapy, where phages can be used to reduce bacterial levels within the chicken to improve poultry health; as disinfectants to decontaminate food processing areas and for food biocontrol to reduce pathogen levels for the consumer (Han *et al.*, 2022). Further discussion on the applications of phage in the poultry industry and examples of experimental studies can be found in review articles (Gigante and Atterbury, 2019; Żbikowska *et al.*, 2020; Thanki *et al.*, 2021).

11.5.1 Commercial phage products and discussion on regulation

Phages have been designated as 'generally regarded as safe' status (GRAS) from the US Food and Drug Administration (FDA) and are considered to be safe. Worldwide, a number of phage products have been approved as processing aids, which in definition is a substance used during the production of food that could end up in the finished product but by law does not need to be disclosed to the consumer as an ingredient (Gambino and Brøndsted, 2021). Table 11.1 summarizes phage products currently on the market, focusing on poultry production, which have already obtained clearance from specific regulatory agencies, such as the FDA and are available to purchase. All approved phages products listed largely focus on reducing bacterial load in food or on food contact surfaces.

Currently there are no phage products that have been approved for direct use in live chickens as either a veterinary medicinal product (VMP) or as a feed additive. A VMP is legally defined as a substance presented as having properties for the treatment or presentation of disease in animals. In comparison, feed additives are products used in animal nutrition with the goal to achieve an effect in the feed itself, in animals, or on food products attained from animals ingesting the feed additive. Within the current regulatory framework, phage products can be regulated as either a VMP or as a feed additive and depends on the claims associated with the product (Thanki *et al.*, 2021). Although there are currently no approved veterinary use phage products, there are a number of products in the pipeline, which are listed in Table 11.2, some of which are awaiting approval in Europe whilst others have been patented and are in development stage.

To our knowledge, the phage product Bafasal® developed by Proteon Pharmaceuticals (Poland) is the furthest along the regulation process; Bafasal® is a four-phage cocktail targeted against common *Salmonella* serotypes associated with chickens and can be delivered via drinking water or feed. In May 2021, EFSA concluded Bafasal® is safe for all avian species and end users. However, the product has not been approved/registered for use yet and is currently awaiting approval by EFSA as a feed additive for avian species.

There is an important requirement for all antibacterial products to undergo a comprehensive safety study and to demonstrate they are safe (not toxic) for the user and this has been conducted for Bafasal®. The study showed Bafasal® does not increase mortality and does not lead to reductions in water or feed intake in birds. In addition, when the phage cocktail was added to feed, it did not cause any visible changes in the smell, structure and appearance of the

Table 11.1. Examples of commercial phage products currently available.

Phage product and manufacturer	Phage or phages in cocktail	Approval status	Notes on product	Reference (all accessed 17 December 2022)
SalmoFresh™ Intralytix Inc. (USA)	Six	FDA-approved as a processing aid, granted GRAS status (GRN000435) Approved by Israel Ministry of Health; Health Canada	• Food treatment on poultry, fish and fruits • Effective against over 900 *Salmonella* strains representing more than 50 serotypes.	https://www.cfsanappsexternal.fda.gov/scripts/fdcc/?set=GRASNotices&id=435&sort=GRN_No&order=DESC&startrow=1&type=basic&search=INTRALYTIX
SalmoPro® Phagelux (Canada)	Two	FDA-approved, granted GRAS status (752) Jul 2018	• For use as an antimicrobial processing aid to control *Salmonella* on food, when applied onto food surfaces	https://www.fda.gov/grasnoticeinventory
Salmonelex™ (Former PhageGuard) Micreos Food Safety BV (The Netherlands)	Two	FDA-approved, granted GRAS status (GRN000630)	• For use as an antimicrobial on foodstuffs to control *Salmonella* • Can be sprayed topically or added to chill tank water	https://www.fda.gov/media/98485/download
EcoShield PX™ Intralytix Inc. (USA)	Three to eight	FDA-approved through a food contact notification, granted GRAS status (GRN000834)	• For use as an antimicrobial on foodstuffs to control *E. coli*	http://www.intralytix.com/files/prod/19EP/19EP-Desc.pdf
Listex™ P100 (Former PhageGuard) Micreos Food Safety BV (The Netherlands)	One	FDA-approved as a processing aid, granted GRAS status	• For use as an antimicrobial on foodstuffs to control *Listeria monocytogenes* • Foodstuffs include ready to eat deli meats, raw seafood and vegetables • Can be used directly on food contact surfaces against *Listeria*	https://phageguard.com/listeria-solution/

Table 11.2. Examples of phage products that have been patented and/or awaiting regulatory approval.

Phage product and/or manufacturer	Phage or phages in cocktail	Notes on product	Reference (all accessed 17 December 2022)
Bafasal® Proteon Pharmaceuticals (Poland)	Four	• Treatment against *Salmonella* in poultry • Can be applied in water or feed • Granted safe status by EFSA and awaiting approval as a feed additive	https://www.proteonpharma.com/bafasal-product/
SalmoFREE® Sciphage (Colombia)	Six	• For treatment and control of *Salmonella* in poultry farms • Phage cocktail can be applied in water or in feed • Patent number WO2017089947A2	https://patents.google.com/patent/WO2017089947A2/en
Biotector CJ Cheiljedang (South Korea)	Unknown	• Phage cocktail with antimicrobial activity against *Salmonella*, *C. perfringens* and *E. coli* • Feed additive for use in poultry and swine • Patent numbers EP3133151, EP3135761, EP2961836, WO2013157813	https://www.cjbio.net/en/products/biotector.do
University of Leicester (United Kingdom)	One to six	• Phage cocktail with antimicrobial activity against *Salmonella* in livestock • Patent number WO2020065302	https://patentscope.wipo.int/search/en/detail.jsf?docId=WO2020065302
ProBe-Bac PE Optipharm (Korea)	One	• Antimicrobial activity against avian pathogenic *E. coli* (APEC) • Can be added to feed or used as a disinfectant or detergent • Patent number KR101830900	http://www.optipharm.co.kr/ENG/ourbusiness/bacteriophage01.php
Synbiolab (Canada)	Two to five	• Antimicrobial feed additive with activity against APEC in poultry • Patent number EP3682001	https://syntbiolab.com/
Purina animal nutrition (USA)	Unknown	• A phage composition to decontaminate feed, which exhibits lytic activity against *Salmonella*, *E. coli*, *Campylobacter* and *Clostridium* for livestock animals including poultry and calves • Patent numbers US20200305461, WO2020205736, CA3133479	https://patents.google.com/patent/WO2020205736A1/en

feed. Furthermore, studies in mice and broilers highlighted the phage product administered at high doses was not toxic, caused no pathological changes in their organs and caused no genotoxicity (mutations in cells) (Wójcik *et al.*, 2020). The dossier submitted to EFSA was reviewed by the panel 'Additive and Products or Substances in Animal Feed' and in May 2021 they concluded Bafasal® is safe for all avian species and end consumers (the opinion of EFSA can be found in Bampidis *et al.*, (2021)). This detailed safety study will considerably add to the body of evidence demonstrating that phages are generally safe and it will provide guidelines/methods on assessing safety of veterinary phage products, which will be needed for the approval of other products.

Proteon Pharmaceuticals included efficacy data on Bafasal® to reduce *Salmonella* colonization in challenged broiler chickens in the dossier submitted to EFSA. They conducted a study with 180 broilers that were divided into three treatment groups: T1 were control birds; T2 birds were challenged with the strain *Salmonella enterica* serovar Enteritidis 12; and T3 birds were challenged and received Bafasal® in water at a dose of 2 x 10^6 plaque forming unit per bird per day. The study lasted for 35 days and, by the end of the trial, birds in group T3 given Bafasal® treatment had significantly reduced *Salmonella* colonization by 16 times in the caeca compared to birds in group T2. Additionally, by day 35, birds in T3 gained more weight, had higher average daily weight gain and their mean daily feed intake was higher in comparison to groups T1 and T2. The feed conversion ratio was also improved in T3 birds, ranging from 1.95, to 2.03 to 1.75 for birds in groups T1, T2 and T3, respectively (Wójcik *et al.*, 2020). Although Bafasal® treatment improved the performance of the birds and reduced *Salmonella* colonization, the EFSA panel was not in a position to conclude the efficacy of Bafasal® for avian species as sufficient data were not provided. Thus, further efficacy studies need to be conducted for EFSA to assess and approve the product as a feed additive for use in birds (Bampidis *et al.*, 2021).

11.6 Impact of Phage Therapy on the Poultry Microbiome

A proposed benefit of phage therapeutics is the precise mode of action compared to broad-spectrum antibiotic therapies (Islam *et al.*, 2021). A wealth of data shows that selective elimination and reduction of pathogens associated with poultry production practices (e.g. *Salmonella* and *Campylobacter*) can be achieved via phage intervention strategies (Mosimann *et al.*, 2021). From a microbial perspective, phage therapeutic applications should be highly specific for the target strain, with minimal interactions with other microorganisms that constitute the host microbiota. To assess phage specificity during a therapeutic treatment, it is possible to analyse the effects on microbial communities via DNA sequencing and metagenomics (shotgun, Illumina short-read sequencing, Nanopore and Pacbio).

Shotgun metagenomics involves the study of environmental DNA and has diverse applications for analysing microbiomes of soil, water, slurry, animals, intestines, etc. Sequencing environmental DNA provides insights into the microbial communities present (the microbiota) and comparisons can be made between, for example, phage treated/untreated samples. Illumina short-read sequencing and Pacbio or Oxford Nanopore long-read sequencing technologies are the platforms typically used to generate metagenomic profiles (Liu *et al.*, 2021; Tedersoo *et al.*, 2021). Another method used to study the microbiome is via amplification of the 16S rRNA gene directly from environmental samples followed by Illumina short-read amplicon sequencing, which can obtain phyla/genus level identification. The 16S rRNA gene is universally distributed in prokaryotes and provides a useful marker owing to the presence of conserved and variable sequences that allow taxonomic identification (Liu *et al.*, 2021). While still limited in terms of published studies, a number of recent investigations have utilized both short-read metagenomics/16S amplicon sequencing to analyse phage/host interactions in poultry production (Thanki *et al.*, 2022a, b).

Broiler houses provide an excellent environment to study the efficacy of phage treatments within the context of the poultry microbiome. The high density of poultry allows for sufficient samples (digesta and faecal) to be obtained for statistical purposes from hatch-to-slaughter, phages can be delivered directly via feed/water systems or directly to birds and the rapid 0–42-day rearing cycle allows for frequent temporal and spatial sampling of various environments (litter/barn floor, faeces, cloaca, caeca, etc.). In addition, for small-scale studies, it is possible to administer phage directly to birds (via oral gavage, feed or water) and monitor individual birds in small pen facilities. However, owing to their complexity and other factors relating to the production cycle, large commercial-scale trials may be negatively influenced by cross-contamination of treatments or the presence of phages in poultry pre-trial (Clavijo *et al.*, 2022). Various phage intervention strategies can therefore be developed and employed; however, data reporting is usually limited to reductions in pathogen levels. Combining culture-based studies with metagenomic approaches provides insights into both the antimicrobial properties of the phage treatment and interactions with the host microbiota.

11.6.1 Microbiome studies on the impact of *Salmonella* phages

A recent study that combined both culture-based and metagenomic methods was conducted on broilers in Columbia, South America. The research group were testing the efficacy of their phage product called SalmoFREE® against *Salmonella* and phage treatment was delivered via water. Two separate field trials were performed at commercial scale in a production facility with ongoing issues of *Salmonella* prevalence with repeated infections of birds housed in the unit. A mixture of Cobb/Ross broilers (trial one n=34,986 and trial two n=34,680) were split into treatment and non-treatment groups housed in separate sheds (note: the broiler breeds were not mixed together in the same house and were placed in different houses). Trial one results were confounded owing to veterinary-recommended antibiotic intervention in SalmoFREE®-treated sheds (Clavijo *et al.*, 2019). However, there was no antibiotic intervention in trial two and a follow-on study of 16S amplicon sequencing in caeca samples collected in trial two indicated that, alongside reducing incidences of *Salmonella*, the use of SalmoFREE® was potentially affecting other members of the poultry microbiota.

Possible modulation of the gut microbiome with SalmoFREE®phages was reported (Clavijo *et al.*, 2022) with interesting observations being made with regards to *Campylobacter* colonization. In this instance it was observed that both treatment groups were colonized with *Campylobacter* at the same time; however, growth rates (as determined by relative abundance) were seemingly suppressed but not abolished in phage-treated groups compared to non-treatment groups. The authors also reported that despite their best efforts to maintain strict biosecurity and separation between experimental treated and untreated groups, components of the SalmoFREE® phage cocktail were found to be present in the untreated sample group presumably due to cross-contamination. The second field trial also revealed the presence of endogenous *Salmonella* phages in pre-treatment samples obtained from the experimental broilers (Clavijo *et al.*, 2022). From an experimental perspective, the issue of cross-contamination between untreated and SalmoFREE®-treated groups may potentially be overcome by increasing biosecurity measures when sampling. However, the presence of endogenous *Salmonella* phages may be influencing results obtained when comparing *Salmonella* CFU g^{-1} caecal contents from SalmoFREE®-treated broilers, with non-treated broilers. Further attempts at large-scale poultry trials encompassing antimicrobial effects of phage treatments and metagenomic or 16S amplicon sequencing approaches should not be discouraged. Refining the design of phage therapeutic poultry trials is always possible and the use of negative results can be used to improve overall goals in the long term (Clavijo *et al.*, 2022).

11.6.2 Microbiome studies on the impact of *Campylobacter* phages

Campylobacter jejuni is the causative agent of campylobacteriosis (usually self-limiting vomiting and diarrhoea symptoms lasting about one week) in humans. Poultry are recognized as an important driving factor in the spread of *C. jejuni* from the food chain to humans. It is estimated that a 2-$\log_{10}$ CFU reduction in *Campylobacter* levels on chicken products at retail has the potential to reduce cases of human campylobacteriosis by up to 30-fold (Loc Carrillo *et al.*, 2005). The application of lytic phages to control *C. jejuni* in broiler chickens and retail meats has been shown on several occasions to be a powerful biocontrol agent. Many studies showed that single phage treatments or phage cocktails were effective at reducing *C. jejuni* levels by up to 2 $\log_{10}$ CFU during *in vitro*, *in vivo* and *ex vivo* (chicken skin/retail meats) experiments.

A recent study investigating the activity of a two-phage cocktail against *C. jejuni* strain HPC5 was combined with 16S rRNA amplicon sequencing (Richards *et al.*, 2019). The phage cocktail was shown to have an initial high-level of suppression of *C. jejuni* HPC5 two days post-treatment with markedly lower caecal levels of HPC5 >2 $\log_{10}$ CFU g^{-1} compared to mock-treated birds. Five days post-phage treatment, HPC5 levels had increased compared to earlier caecal counts, but were still >1.3 $\log_{10}$ CFU g^{-1} lower than in the control group. To ascertain whether the administration of *Campylobacter*-specific phages has any effects on the poultry microbiota, the authors performed 16S amplicon sequencing on samples obtained throughout the experiment. The first observation was that caecal and ileal microbial diversity between phage-treated and mock-treated birds did not differ significantly throughout the experiment. However, at day five, caecal species richness was observed to differ significantly in phage-treated groups when compared to non-treatment groups. In phage-treated samples, it was apparent that the relative abundance of *Campylobacter* had been greatly reduced compared to mock-treated chickens (Richards *et al.*, 2019).

11.6.3 Discussion on the future of microbiome phage studies

With only a handful of poultry-associated trials coupling metagenomic or 16S amplicon sequencing to phage therapeutic data, such studies represent a research area that is vastly under-investigated. The lack of data in the literature is largely associated with previously cost-inhibitive prices associated with obtaining metagenome sequences. It is apparent from the above-mentioned analyses of SalmoFREE® (Clavijo *et al.*, 2022) and the efficacy of *Campylobacter* phages in broiler chickens (Richards *et al.*, 2019) that potentially beneficial changes in microbial population structure can be delivered via phage intervention strategies in poultry. Metagenomics and 16S amplicon sequencing allows unprecedented insights into phage/bacterial interactions. As more studies are performed looking at the potential benefits of phages in controlling pathogens of poultry, it will be interesting to see how phage treatments and metagenomics can be linked to growth performance of birds.

Engineering poultry microbiomes to optimize the colonization of beneficial bacteria may be possible with targeted phage intervention strategies. A number of studies and reviews can be found in the literature that have directly investigated the poultry microbiome in association with population structure, AMR genes and pro/prebiotics (Wright *et al.*, 1994; Kogut, 2019; Ricke *et al.*, 2020; Shehata *et al.*, 2022). An interesting approach from Gupta *et al.* (2021a) investigated the collection of antibiotic resistance genes (ARGs) in used poultry barn litter using a spatial and temporal approach to sampling. Broiler litter is often used as a fertilizer owing to its high concentrations of nitrogen, phosphorous, etc. The authors observed differences in the communities of several pathogens of clinical and veterinary health concerns when comparing the top surface, middle layers and bottom layer of the litter. Over the time course of the experiment, shifts in these communities were also observed, allowing assumptions to be made as to the best approaches for waste management prior to application as a fertilizer (Gupta *et al.*, 2021b).

In a parallel study, they also assessed the differences in poultry-associated microbiomes obtained from cloacal swabs when compared with litter microbiomes. The authors conclude that the two sites of sampling were significantly different such that within poultry production environments cloacal swabs represent the most suitable approach for source tracking of disease outbreaks, whereas litter sampling would be more suited for assessing environmental routes of transmission, e.g. crop fertilization and agricultural run-off (Gupta *et al.*, 2021a). Agricultural by-products such as poultry litter may present public/ animal health risks owing to the potential presence of ARGs and pathogens. Investigating the role of phages in the poultry production environment for purposes of safe waste-management may lead to better practices being implemented.

11.6.4 Studying the poultry virome

To complement metagenomic datasets relating to the bacteria/archaeal/fungal members of the microbiota, analysis of the poultry virome may be performed. The poultry virome consists of animal viruses (e.g. *Picornaviridae*, *Reoviridae*, *Adenoviridae*, *Caliciviridae* and *Picobirnaviridae*) and phages. To date no studies specifically focused on the phage fraction of a poultry metagenome have been performed and no studies have investigated if phage therapy alters the poultry virome. However, Day *et al.* (2015) performed an in-depth analysis of the poultry gut microbiome and RNA virome and reported the presence of a number of sequences related to *Siphoviridae*, *Podoviridae* and *Myoviridae* phages. From hatching through to processing, metagenomic study of the poultry virome was used to highlight the presence or absence of a number of phage species infecting various *Enterobacteriaceae* (e.g. *E. coli*, *Salmonella* spp., *Shigella* spp.) during the grow out cycle (Mulholland *et al.*, 2021). The study revealed strong correlations (using Pearson's coefficient of correlation) between the presence of various phage types at different anatomical locations. At hatching, the broiler trachea was observed to harbour podoviruses infecting *Pseudomonas* spp., *Brevibacterium* spp, *Flavobacterium* spp. and *Streptococcus* spp. Over time, shifts in phage and bacterial communities showed siphoviruses and podoviruses capable of infecting *Dermabacteraceae* were found to be strongly correlated with the presence of the host. Furthermore, common inhabitants of the avian trachea throughout the rearing cycle are *Staphylococcus* spp. and accordingly a strong correlation between the presence of host and *Staphylococcus*-infecting siphoviruses were observed from start to finish (Mulholland *et al.*, 2021).

Multiple studies have been undertaken to analyse viral communities, including the eukaryotic faecal virome of healthy broilers (Lima *et al.*, 2017), viromes of broilers with active respiratory tract infections (Sajnani *et al.*, 2018) and the previously mentioned intestinal microbiota and RNA viromes of 'sentinel' birds placed on commercial broiler farms (Day *et al.*, 2015). From the literature, it is apparent that there is a knowledge gap in the role of phages in the poultry virome. While this shortfall currently exists, it is anticipated that more research focus will be placed on studying the poultry virome, especially within the context of phages and how they interact with their bacterial hosts and the impact phage therapy has on the virome.

11.7 Advantages and Limitations (with Solutions) of Phage Therapy

Antibiotics have a broad spectrum of activity and target multiple bacterial species, including beneficial pathogens. In comparison, phages are bacterial species specific, which avoids disruption of the microbiota and elimination of potentially beneficial bacteria. This helps to avoid overgrowth of other potentially harmful pathogens and growth of multi-drug resistant bacteria (Huang *et al.*, 2022). In addition, the action of phages is limited to the site of infection, instead of being widespread across the whole human or animal body. Furthermore, phages are safe as they cannot infect human or animal cells. Phages are well tolerated as both humans and animals are continuously exposed to them and therefore by using phages therapeutically a new biological entity is not being introduced into the system. The safety of phage therapy has been confirmed in knowledge gathered in Eastern Europe and experimental phage therapy studies, which have all reported no adverse side effects of phage therapy (Żaczek *et al.*, 2020; Chanishvili *et al.*, 2022).

Potential limitations of phage therapy include isolation of therapeutic phages as, for example, it is more straight forward to isolate *E. coli* phages then to isolate phages against *Brachyspira hyodysenteriae*, the causative agent of swine dysentery. This is due to a wealth of information available for *E. coli* such as protocols to grow the bacterium, to isolate phages and to study phage-bacterial interactions, whilst in comparison there are limited scientific studies on *B. hyodysenteriae* and its phages as its difficult to grow in the laboratory (Huang *et al.*, 2022). A second limitation is that phages are pH sensitive, and the pH in the chicken ranges from 1 to 7 from the crop to caecum, which could potentially inactivate pH-sensitive phages. However, multiple *in vivo* studies have shown this is not the case and phages administered to chickens have been reisolated in faecal and caecal samples (Thanki *et al.*, 2021). If phage stability is an issue, there are methods available to formulate and stabilize phages and this is a growing area of research within the phage community (Malik *et al.*, 2017; Malik and Resch, 2020).

Similar to antibiotics, there is a possibility of bacteria becoming resistant to phages, as bacteria can develop various mechanisms to avoid phage infection. To overcome this, phage cocktails are used therapeutically and the cocktail includes genetically diverse phages, which bind to different receptors on the bacterial cell wall/membrane. Phages typically bind to proteins on the surface of the bacterial cells, which play a vital role in the function of the cell and consequently mutating the receptor sites induces a fitness cost to the bacterium. The costs include: the bacteria becomes less virulent, is unable to colonize as efficiently as before and studies suggest phage-resistant bacteria show increased susceptibility to antibiotics (Mangalea and Duerkop, 2020). Consequently, research groups are currently investigating if phage and antibiotics can be used in synergy (Tagliaferri *et al.*, 2019).

Currently, there is limited research on the interaction of phage with the immune system but published research to date suggests a positive synergy between the systems. One such research study investigated the effect of dietary supplement of phage in *Salmonella*-challenged birds on their immune systems. The groups showed phage treatment did lead to an immune response, which was a positive up regulation of pro and anti-inflammatory cytokines. Both responses play an important role in antimicrobial host defences and the interaction between phages and the immune response probably led to significant reductions in *Salmonella* colonization within the ileum (Adhikari *et al.*, 2017).

Lastly, a potential hurdle could be scaling up production of phages to meet commercial needs. To overcome this obstacle, researchers are investigating phage production using bioreactors where 1 to 156 litres of phages at high concentrations can be produced (García *et al.*, 2019). Bioreactors are vessels or tanks that have been designed to provide an environment to transform biochemicals into products and are commonly used by the poultry industry to produce pre and probiotics. Using bioreactors will be the most cost-effective process and initial analysis suggests the whole production cost from using a 156-litre capacity bioreactor to the filtration step to separate the phage from bacteria could potentially cost US$0.02 per chicken (Torres-Acosta *et al.*, 2019). In addition, there are private companies that specialize in large-scale phage production, such as ACD Pharma in Norway (https://acdpharma.com/en/about-acd-pharma/) and Jafral in Slovenia (https://jafral.com/), that have the capacity to produce phages against multiple pathogens.

11.8 Conclusion

Industrial awareness is growing on the urgent need to reduce antibiotic usage in animal production to limit the spread of multi-drug resistant bacterial strains within the food chain, which is motivating the industry and government bodies to invest in research into alternative antimicrobials. Phages provide a natural alternative to antibiotics to reduce levels of pathogenic bacteria within the poultry food chain. Multiple experimental efficacy studies have provided evidence on the success of phage therapy in poultry, emphasized phages are safe and shown phage treatment can help to improve feed conversion ratios. Furthermore, microbiome studies are highlighting phages are having a positive impact on the microbiome and, with recent advances in genomics plus the cost of sequencing reducing, it will allow for further in-depth microbiome studies to be conducted. Whilst there are food safety phage products being sold worldwide with proven efficacy, there are currently no phage products approved for direct use in live chickens, although many phage products have been patented and one product to our knowledge is undergoing the regulatory process in the EU. Once one veterinary phage product has been approved, it will pave the way for many others and will provide a regulatory framework to follow.

References

Ackermann, H.W. (2009) Phage classification and characterization. In: *Bacteriophages: Methods and protocols*, pp. 127–140. doi: 10.1007/978-1-60327-164-6.

Adhikari, P.A. Cosby, D.E., Cox, N.A., Lee, J.H. and Kim, W.K. (2017) Effect of dietary bacteriophage supplementation on internal organs, fecal excretion, and ileal immune response in laying hens challenged by *Salmonella* Enteritidis. *Poultry Science* 96(9), 3264–3271. doi: 10.3382/ps/pex109.

Adriaenssens, E.M., Sullivan, M.B.,Knezivic, P., van Zyl, L.J., Sarkar, B.L. *et al.* (2020) Taxonomy of prokaryotic viruses: 2018-2019 update from the ICTV Bacterial and Archaeal Viruses Subcommittee. *Archives of Virology* 165(5), 1253–1260. doi: 10.1007/s00705-020-04577-8.

Bampidis, V., Azimonti, G., de Lourdes Bastos, V., Christensen, H., Dusemund, B. *et al.* (2021) Safety and efficacy of a feed additive consisting on the bacteriophages PCM F/00069, PCM F/00070, PCM F/00071 and PCM F/00097 infecting *Salmonella* Gallinarum B/00111 (Bafasal®) for all avian species (Proteon Pharmaceuticals S.A.). *EFSA Journal* 19(5). doi: 10.2903/j.efsa.2021.6534.

Chanishvili, N., Myelnikov, D. and Blauvelt, T.K. (2022) Professor Giorgi Eliava and the Eliava Institute of Bacteriophage. *PHAGE: Therapy, Applications, and Research* 3(2), 71–80. doi: 10.1089/phage.2022.0016.

Clavijo, V. Baquero, D., Hernandez, S., Farfan, J.C., Arias, J. *et al.* (2019) Phage cocktail SalmoFREE® reduces *Salmonella* on a commercial broiler farm. *Poultry Science*, 98(10), 5054–5063. doi: 10.3382/ps/pez251.

Clavijo, V., Morales, T., Vives-Flores, M.J. and Munoz, A.R. (2022) The gut microbiota of chickens in a commercial farm treated with a *Salmonella* phage cocktail. *Scientific Reports* 12(1), 1–16. doi: 10.1038/s41598-021-04679-6.

Clokie, M.R.J., Millard, A.D., Letarov, A.V. and Heaphy, S. (2011) Phages in nature. *Bacteriophage* 1(1), 31–45. doi: 10.4161/bact.1.1.14942.

Day, J. M. Oakley, B.B., Seal, B.S. and Zsak, L. (2015) Comparative analysis of the intestinal bacterial and RNA viral communities from sentinel birds placed on selected broiler chicken farms. *PLoS ONE* 10(1), 1–15. doi: 10.1371/journal.pone.0117210.

EFSA and ECDC (2022) The European Union summary report on antimicrobial resistance in zoonotic and indicator bacteria from humans, animals and food in 2019–2020. *EFSA Journal* 20(3). doi: 10.2903/j.efsa.2022.7209.

Gadde, U., Kim, W.H., Oh, S.T. and Lillehoj, H.S. (2017) Alternatives to antibiotics for maximizing growth performance and feed efficiency in poultry: a review. *Animal Health Research Reviews* 18(1), 26–45. doi: 10.1017/S1466252316000207.

Gambino, M. and Brøndsted, L. (2021) Looking into the future of phage-based control of zoonotic pathogens in food and animal production *Current Opinion in Biotechnology* 68, 96–103. doi: 10.1016/j.copbio.2020.10.003.

García, R., Latz, S., Romero, J., Higuera, G., Garcia, K. *et al.* (2019) Bacteriophage production models: an overview. *Frontiers in Microbiology* 10, 1–7. doi: 10.3389/fmicb.2019.01187.

Gigante, A. and Atterbury, R.J. (2019) Veterinary use of bacteriophage therapy in intensively-reared livestock. *Virology Journal* 16(1), 155. doi: 10.1186/s12985-019-1260-3.

Gupta, C.L., Blum, S.E., Kattusamy, K., Daniel, T., Druyan, S. *et al.* (2021a) Longitudinal study on the effects of growth-promoting and therapeutic antibiotics on the dynamics of chicken cloacal and litter microbiomes and resistomes. *Microbiome* 9(1), 1–19. doi: 10.1186/s40168-021-01136-4.

Gupta, C. L., Avidov, R., Kattusamy, K., Saadi, I., Varma, V.S. *et al.* (2021b) Spatial and temporal dynamics of microbiomes and resistomes in broiler litter stockpiles. *Computational and Structural Biotechnology Journal* 19, 6201–6211. doi: 10.1016/j.csbj.2021.11.020.

Han, S., Byun, K.H., Mizan, M.F.R., Kang, I. and Ha, S.D. (2022) Bacteriophage and their lysins: a new era of biocontrol for inactivation of pathogenic bacteria in poultry processing and production—A review. *Food Control* 137, 108976. doi: 10.1016/j.foodcont.2022.108976.

Hedman, H.D., Vasco, K.A. and Zhang, L. (2020) A review of antimicrobial resistance in poultry farming within low-resource settings. *Animals* 10(8), 1–39. doi: 10.3390/ani10081264.
Huang, Y., Wang, W., Zhang, Z., Gu, Y., Huang, A. *et al.* (2022) Phage Products for Fighting Antimicrobial Resistance. *Microorganisms* 10(7), 1324 doi: 10.3390/microorganisms10071324.
Islam, M.R., Martinez-Soto, C.E., Lin, J.T., Khursigara, C.M., Barbut, S. *et al.* (2021) A systematic review from basics to omics on bacteriophage applications in poultry production and processing. *Critical Reviews in Food Science and Nutrition* 0(0), 1–33. doi: 10.1080/10408398.2021.1984200.
Jäckel, C., Hammerl, J.A. and Hertwig, S. (2019) *Campylobacter* phage isolation and characterization: what we have learned so far. *Methods and Protocols* 2(1), 1–11. doi: 10.3390/mps2010018.
Kogut, M.H. (2019) The effect of microbiome modulation on the intestinal health of poultry. *Animal Feed Science and Technology* 250, 32–40. doi: 10.1016/j.anifeedsci.2018.10.008.
Lima, D.A., Cibulski, S.P., Finkler, F., Teixeira, T.F., Varela, A.P.M. *et al.* (2017) Faecal virome of healthy chickens reveals a large diversity of the eukaryote viral community, including novel circular ssDNA viruses. *Journal of General Virology* 98(4), 690–703. doi: 10.1099/jgv.0.000711.
Liu, Y.X., Qin, Y., Chen, T., Lu, M., Qian, X. *et al.* (2021) A practical guide to amplicon and metagenomic analysis of microbiome data. *Protein and Cell* 12(5), 315–330. doi: 10.1007/s13238-020-00724-8.
Loc Carrillo, C., Atterbury, R.J., El-Shibiny, A., Connerton, P.L., Dillon, E. *et al.* (2005) Bacteriophage therapy to reduce *Campylobacter jejuni* colonization of broiler chickens. *Applied and Environmental Microbiology* 71(11), 6554–6563. doi: 10.1128/AEM.71.11.6554-6563.2005.
Malik, D.J. and Resch, G. (2020) Editorial: manufacturing, formulation and delivery Issues for phage therapy to become a reality. *Frontiers in Microbiology* 11, 3389–3391. doi: 10.3389/fmicb.2020.584137.
Malik, D.J., Sokolov, I.J., Vinner, G.K., Mancuso, F., Cinquerrui, S. *et al.* (2017) Formulation, stabilisation and encapsulation of bacteriophage for phage therapy. *Advances in Colloid and Interface Science* 249, pp. 100–133. doi: 10.1016/j.cis.2017.05.014.
Mangalea, M.R. and Duerkop, B.A. (2020) Fitness trade-offs resulting from bacteriophage resistance. *American Society for Microbiology* 88, 1–15. doi: 10.1128/IAI.00926-19.
Miller, R.W., Skinner, E.J., Sulakvelidze, A., Mathis, G.F. and Hofacre, C.L. (2010) Bacteriophage therapy for control of necrotic enteritis of broiler chickens experimentally infected with *Clostridium perfringens. Avian Diseases Digest* 5(1), e28–e29. doi: 10.1637/9256.1.
Mosimann, S., Desiree, K. and Ebner, P. (2021) Efficacy of phage therapy in poultry: a systematic review and meta-analysis. *Poultry Science,* 100(12), 101472. doi: 10.1016/j.psj.2021.101472.
Mulholland, K.A., Robinson, M.G., Keeler, S.J., Johnson, T.J., Weber, B.W. *et al.* (2021) Metagenomic analysis of the respiratory microbiome of a broiler flock from hatching to processing. *Microorganisms* 9(4), 1–18. doi: 10.3390/microorganisms9040721.
Murray, C.J., Ikuta, K.S., Sharara, F., Swetschinski, L., Aguilar, G.R. *et al.* (2022) Global burden of bacterial antimicrobial resistance in 2019: a systematic analysis. *The Lancet* 399(10325), 629–655. doi: 10.1016/S0140-6736(21)02724-0.
Nale, J. Y., Spencer, J., Hargreaves, K.R., Buckley, A.M., Trzepiński, P. *et al.* (2016) Bacteriophage combinations significantly reduce *Clostridium difficile* growth *in vitro* and proliferation *in vivo. Antimicrobial Agents and Chemotherapy* 60(2), 968–981. doi: 10.1128/AAC.01774-15.
Nale, J.Y., Spencer, J., Hargreaves, K.R., Buckley, A.M., Trzepinski, P. *et al.* (2021) 'An optimized bacteriophage cocktail can effectively control *Salmonella in vitro* and in *Galleria mellonella*', *Frontiers in Microbiology* 11(January). doi: 10.3389/fmicb.2020.609955.

O'Neill, J. (2016) Tackling drug-resistant infections globally: final report and recommendations. (May). doi: 10.1016/j.jpha.2015.11.005.

Richards, P.J., Connerton, P.L. and Connerton, I.F. (2019) Phage biocontrol of *Campylobacter jejuni* in chickens does not produce collateral effects on the gut microbiota. *Frontiers in Microbiology* 10, 1–10. doi: 10.3389/fmicb.2019.00476.

Ricke, S.C., Lee, S.I., Kim, S.A., Park, S.H. and Shi, Z. (2020) Prebiotics and the poultry gastrointestinal tract microbiome. *Poultry Science,* 99(2), 670–677. doi: 10.1016/j.psj.2019.12.018.

Sajnani, M.R., Sudarsanam, D., Pandit, R.J., Oza, T., Hinsu, A.T. *et al.* (2018) Metagenomic data of DNA viruses of poultry affected with respiratory tract infection. *Data Brief* 16, 157–160. doi: 10.1016/j.dib.2017.11.033.

Shehata, A.A., Yalcin, S., Latorre, J.D., Basiouni, S., Attia, Y.A. *et al.* (2022) Probiotics, prebiotics, and phytogenic substances for optimizing gut health in poultry. *Microorganisms* 10(2), 1–34. doi: 10.3390/microorganisms10020395.

Shen, A. and Millard, A. (2021) Phage genome annotation: where to begin and end. *PHAGE: Therapy, Applications, and Research* 2(4), 183–193. doi: 10.1089/phage.2021.0015.

Tagliaferri, T.L., Jansen, M. and Horz, H.P. (2019) Fighting pathogenic bacteria on two fronts: phages and antibiotics as combined strategy. *Frontiers in Cellular and Infection Microbiology* 9, 22. doi: 10.3389/fcimb.2019.00022.

Tedersoo, L., Albertsen, M. and Anslan, S. (2021) Perspectives and benefits of high-throughput long-read. *Applied Environmental Microbiology* 87(17), 1–19.

Thanki, A.M., Brown, N., Millard, A.D. and Clokie, M.R.J. (2019) Genomic characterization of jumbo *Salmonella* phages that effectively target United Kingdom *Salmonella* serotypes. *Frontiers in Microbiology* 10, 1–14. doi: 10.3389/fmicb.2019.01491.

Thanki, A.M., Hooton, S., Gigante, A.M., Atterbury, R.J. and Clokie, M. (2021) *Potential roles for bacteriophages in reducing Salmonella from poultry and swine. IntechOpen.* Available at: https://www.intechopen.com/books/advanced-biometric-technologies/liveness-detection-in-biometrics (accessed 6 April 2023).

Thanki, A.M., Clavijo, V., Healy, K., Wilkinson, R.C., Sicheritz-Pontin, T. *et al.* (2022a) Development of a phage cocktail to target *Salmonella* strains associated with swine. *Pharmaceuticals* 15(1).

Thanki, A.M., Mignard, G., Atterbury, R.J., Barrow, P., Millard, A.D. *et al.* (2022b) Prophylactic delivery of a bacteriophage cocktail in feed significantly reduces *Salmonella* colonization in pigs. *Microbiology Spectrum* 10(3). doi: 10.1128/spectrum.00422-22.

Torres-Acosta, M.A., Clavijo, V., Vaglio, C., Gonzalez-Barrios, A.F., Vivez-Florez, M.J. *et al.* (2019) Economic evaluation of the development of a phage therapy product for the control of *Salmonella* in poultry. *Biotechnology Progress* 35(5), 1–11. doi: 10.1002/btpr.2852.

Wójcik, E.A., Stanczyk, M., Wojtasik, A., Kowalska, J.D., Nowakowska, M. *et al.* (2020) Comprehensive evaluation of the safety and efficacy of BAFASAL® bacteriophage preparation for the reduction of *Salmonella* in the food chain. *Viruses* 12(7). doi: 10.3390/v12070742.

Wommack, K.E., Williamson, K.E., Helton, R.R., Bench, S.R. and Winget, D.M. (2009) Methods for the isolation of viruses from environmental samples. *Methods in Molecular Biology* 501, 3–14. doi: 10.1007/978-1-60327-164-6_1.

Wright, D.J., Chapman, P.A. and Siddons, C.A. (1994) Immunomagnetic separation as a sensitive method for isolating *Escherichia coli* O157 from food samples. *Epidemiology and Infection* 113(1), 31–39.

Żaczek, M., Weber-Dabrowska, B., Miedzybrodski, R., Lusiak-Szelachowska, M. and Gorski, A. (2020) Phage therapy in Poland – a centennial journey to the first ethically approved rreatment facility in Europe. *Frontiers in Microbiology* 11. doi: 10.3389/fmicb.2020.01056.

Żbikowska, K., Michalczuk, M. and Dolka, B. (2020) The use of bacteriophages in the poultry industry. *Animals* 10(5). doi: 10.3390/ani10050872.

CHAPTER 12
Phytotherapeutics: Practical Use and Field Experience

Allan Ball*

Slate Hall Veterinary Practice, Lincoln, UK

Abstract

Phytotherapeutics are defined as medicines derived from plants or herbs that are used to treat or prevent health conditions. Their use in intensive poultry production systems has increased significantly in recent years as producers have searched the myriad of available supplementary products to optimize health and performance of their flocks. Choosing the right product for the targeted issue can be challenging for the veterinarian or producer, due to the high number of variables at play and the lack of scientific evidence on the subject. Intensively produced birds will be confronted with environmental and health challenges throughout the production phases. The increased use of phytotherapeutic products is, at least in part, due to the focus that the poultry industry has shown in taking a leading role in reducing the use of antibiotics across their agricultural operations. Annual reports published by the British Poultry Council (BPC) highlight the reduction in antibiotic use in poultry farming from 2012 to 2021. While this has been achieved through renewed focus on animal-production requirements and an improved understanding of the genetic potential of the modern-day chicken, the use of supplementary products such as phytotherapeutics have played an important role. Although the currently available scientific data relating to the efficacy of this classification of products remain low, the field experience that is being gained from their use is increasing rapidly. It has become evident that phytotherapeutic use should not be seen as a direct substitute for antibiotic interventions; their positive impact on production seems to be far greater when used in a preventative model rather than in a typical reactive-type approach. The choice of product used is highly dependent on which organ system is being targeted and, as such, the formulation is fundamental

*Corresponding author: allan.ball@slatehall.co.uk

DOI: 10.1079/9781800622746.0012

in the efficacy of the product. As the number of supplementary options available in this field increase, the challenge on the farm has shifted to field-staff training and developing diagnostic tools that allow sufficiently early identification of abnormalities to allow sufficient time for intervention strategies to have maximum effect on outcome.

12.1 Introduction

Phytotherapy is defined as the use of plants or plant products for medicinal use and therapeutic is defined as medicine concerned with the treatment of disease and the action of remedial agents. Phytotherapeutics can be described as a broad subset of plant-derived bioactive compounds. The products derived from plant parts, specifically essential oils, are known to possess active ingredients that exhibit antimicrobial activity against bacteria, yeast and moulds. The use of phytotherapeutic products in the poultry industry over the past few years has increased significantly, mainly due to the pressures that that industry has faced to reduce its antibiotic usage. They can be added to the diets or supplemented through the drinking water systems of commercial animals to improve their productivity through enhancing feed properties, promoting animals' production performance and improving the quality of products derived from these animals. Among the major groups of principle ingredients that impart antimicrobial properties in their essential oils are thymol, saponins, flavonoids, carvacrol, terpenes and their precursors. This chapter is presented from the perspective of veterinary professionals and relates to their experience in practice.

12.2 Modern-day Production

The balance in the modern-day production system between the states of healthy and diseased is a very delicate one that needs optimal input from the farmer through management, the bird through physiological optimization and the facilities through providing optimal conditions for growth and performance, not forgetting the impact that biosecurity must play in the whole cycle. Biosecurity acts as the pillar of disease prevention in intensive production systems, but other factors such as vaccination also have their part to play in disease prevention. As our production systems strive for further improvements in performance, so the balance tipping point is that much more sensitive to changes in either of the above-mentioned key contributors. Our farmers, stockpersons and veterinarians are required to identify abnormalities earlier to ensure that the use of phytotherapeutics stand a chance in the face of daily challenges, as their efficacy is enhanced through earlier intervention strategies on farms.

The increased focus on and use of phytotherapeutic as alternatives to antibiotic intervention has in part, been as a result of the British Poultry Council Antibiotic Stewardship programme, launched in 2012, which challenged its member to reduce their antibiotic use across their operations. This action has challenged the industry to better understand alternative production and treatment options to enhance the wellbeing and performance of the bird. Banning of the use of antibiotic growth promoters in livestock in January 2006 (EU Regulation (EC) No 1831/2003) put additional health pressures on farming enterprises and at the time alternative therapy options were not deemed as appropriate interventions, or indeed necessary. This, however, changed and has gathered momentum through the years and continues to do so. It is fundamental that we understand the limitations of the modern broiler and layer to perform at their highest levels and to know that the organ systems are challenged continuously to function at their highest potential to achieve the breed performance targets.

12.3 Programme Implementation: Key Decisions

Implementing an effective phytotherapeutic strategy begins like any other veterinary investigation. A full farm and flock history are the foundation as without these there is no backbone to any intervention option. Farming, and especially intensive livestock farming, is blessed with exceptional record keeping and flock-monitoring records. These records enable the farmer, producer and veterinarian to gather and interpret data, and to accurately highlight abnormalities and deviations from the breed targets that serve as an early indicator or warning sign that the animal health is moving from optimal to sub-optimal, with this being exhibited as small changes in drinking water intake, feed consumption or changes to the shed environment, such as a rise in shed humidity or ammonia level. Along with a review of the available data, the stockperson also serves as a good source of information if they are able to identify the early signs indicative of sub-optimal physiological function. With their help in the forefront of the day-to-day review of the farm health status, as well as a review of the available data, this serves as an excellent point to identify potential deviations from expected breed targets. With exceptional bird performance also comes the need to provide the necessary conditions to enable the bird to express their full genetic potential. This is gained through providing the necessary environmental conditions to maintain birds within their thermo-neutral zone, and to reduce environmental challenges through optimal health and hygiene of the environment, which starts when the farm is undergoing its terminal cleaning and disinfection between cycles. Through the growing cycle control of the air quality and litter quality are paramount; however, it is the period between crops when there are no birds on the farm that the true reduction in viral and bacterial loading can be achieved through a well-managed and efficient cleaning and disinfection programme. Another important part of the history involves collection and analysis of previous flock health history through review of records from further diagnostic work completed, to include, but not limited to, post-mortem examination, additional field diagnostic testing (bacteriology, serology, histology, PCR testing) and to review results from routine flock surveillance, which is carried out to monitor background health status.

To ensure desired outcome measures we need to go back to the basics and consider the anatomy and the physiological function of the target organ system. For example, the intestinal tract of poultry is, in comparison to other species, very short but has functional adaptations to optimize its function. One of these is the ability to exhibit both peristalsis and retroperistalsis (the reverse of the involuntary smooth muscle contractions of peristalsis) to optimize feed digestibility and nutrient absorption. If any of the normal physiological function is compromised, this soon becomes apparent when examining the bird and by the output into the environment through evidence of an abnormal faecal mass and wetter litter in the sheds. To understand what the normal function is and what this looks like is fundamental, as early identification of the abnormal intestinal function will prove vital in ensuring the interventions with phytotherapeutic products can have their desired effect. On this note, it is also beneficial to train and inform the farm staff on the difference between normal and abnormal with regards to faecal examination and for this information to be reported back to the veterinarian. Sub-optimal function that can compromise the overall intestinal health in cases of a high coccidiosis challenge, dysbiosis/dysbacteriosis, necrotic enteritis and infectious runting and stunting signify the breakdown in the normal intestinal functionality, which means that the intestinal health is compromised. If normal physiological function is compromised, it is highly likely that the use of phytotherapeutic products will be below expectation. When reviewing the normal physiological function of the intestinal tract, the main reason for breakdown in the intestinal integrity is through adverse environmental, hygiene or disease factors. These ultimately have the impact of destabilizing the function of the tight junctions in the intestinal tract, resulting in these junctions becoming 'leaky', allowing the through-flow of intestinal bacterial commensals from the intestinal tract into the generalized circulation. The result of this is a state of sepsis. A number of phytotherapeutic products have anti-inflammatory effects which could help maintain the tight junction integrity when birds are faced with the disruption of the normal intestinal function.

The following step is then to decide on what your appropriate product choice shall be, depending on your desired outcome and what target organ you are focused on. Take into consideration the product formulation, recommended dosing regime, farm compliance with regards to dosing pumps and farm manager understanding of the process and finally how to monitor its progress throughout the dosing period. There are significant differences in the available formulations of products and a significant amount of time and money has gone into field trials to determine which of these is suitable to use. Our experience as veterinary professionals has shown that in many cases supplementation with phytotherapeutics alone for supporting intestinal health may not exhibit the fully expected outcome, but their use for intestinal health cases can be enhanced and amplified when combined with other products such as prebiotics, probiotics and intestinal acidifiers. The more effective intestinal acidifier products are weak acids that have a better impact in the lower intestinal tract (ileum and caeca) where the unfavourable bacteria such as *Escherichia coli*, *Enterococcus* and *Salmonella* inhabit.

When the recommended programme is followed and the farm is not challenged with other bacterial or viral challenges associated with intensive farming practice, the outcome can be very rewarding. An example of when maximal impact can be achieved through dosing is exhibited when we focus our supplementation strategies into the period of maximal intestinal development and maximum intestinal instability, namely the first 0–7 days of the bird's life on the farm. During this period the intestinal tract is developing both in length (day 0–2) and undergoing cell differentiation (day 3–7) (Jin *et al.*, 1998), as well as having an unbalanced intestinal microflora as a result of the mixing of bacterial population passed down from the parent, combining with the farms established microflora. This results in a period of intestinal health instability, which can be supported through assisting to accelerate the development of mature microflora and achieve a closer state of stability sooner (Kogut, 2019). When this period is optimized through satisfactory nutritional input, and additional management focus is placed on the brooding period, supplementation with phytotherapeutic products can significantly improve intestinal development through improved villus height, increasing the intestinal absorptive surface area and reducing the harmful impact of the unfavourable intestinal bacterial populations. This will show its benefit later in the broiler crop with drier litter conditions, as well as positive impacts on feed conversion.

Supplementation strategies through the drinking water in general follow a 3–5-day duration, which is repeated on a weekly basis if required. Supplementation through in-feed additives will depend on feed bin capacity on the farm, the amount that the birds are eating on a daily basis and how many feed deliveries are scheduled for the farm. Both administration methods have their pros and cons. Water-based programmes can be implemented rapidly and can be controlled as a dip-in, dip-out strategy when flocks are going through acute stress periods. However, they can have adverse effect on the drinker systems of the houses and accurate dosing is dependent on how well the incoming water-dosing systems are managed. However, they are generally cheaper than in-feed supplementation for the specific reason that the duration of administration can be short and targeted. In-feed programmes offer supplementation for longer periods, for as long as birds have access to feed, but are usually slow to implement and can be more costly. However, the farm manager does not need to intervene in any way as feed is readily available to the birds.

12.4 Product Choice and Field Implementation: Intestinal Health

The overall choice of phytotherapeutic options that have become available over the last five years has increased significantly. However, there is still a genuine lack of published data to support a lot of their uses. Compared to data published in the human health sector on supplements and essential oils use, the data available to support phytotherapeutic impact in farming is comparatively sparse. This has resulted partly in significant product choice into the market, but for the farmer and veterinarian has created a headache, as all products seemingly have a claim to improvement in intestinal health, feed conversion and mortality. Our alternative health solutions used to be a simple 1-2-3 approach; we are now faced with putting together complex programmes for farms to follow. When successful, however, they prove to add a significant cost benefit to the farming enterprise. With increased phytotherapeutic product choice also comes complications with product compliance and use to ensure these products are dosed accurately and correctly and that timing during the production cycle is optimal to ensure their efficacy is optimized. Within all these parameters the products must also meet Hazard Analysis and Critical Control Point (HACCP) and Good Manufacturing Process (GMP) production criteria to ensure satisfactory product quality. This also ensures that the essential oil products do not leave adverse residues or result in final product tainting which would result in consumer complaints. Our own experience has shown that product performance needs to be reviewed in field conditions before we include these into our portfolio. Once the tailored programme for the farm has been made, the production team need correct guidance and education into the factors that make these interventions successful: first, the product mode of action: second, the timing of administration; and third, to ensure accurate delivery to the birds.

The additional requirements from our local UK retail base to source locally produced products have also aided a more focused approach to alternative intervention strategies as they too focus their efforts on reduction of antibiotic use as a key performance indicator (KPI). As more and more pressure is placed on intensive agricultural systems to reduce their antibiotic use, so the ongoing discussion about antibiotic resistance continues to rumble in the background. The number of publications on this topic increases annually and so our efforts as an industry to continue our valiant progress continues.

Within the UK poultry industry, our most common reason for the use of antibiotic products remains the clinical presentation of enteritis/dysbacteriosis. The main driving forces behind this are the physiological limitations of the chicken's gastrointestinal tract, as well as the pressure to reduce the cost of feed rations and feed lower density rations. Seeing as feed in the broiler growing cycle accounts for over 70% of the production costs, the profitability margin can be driven primarily through optimization of the farms' feed conversion. As noted previously, however, the short intestinal tract and hence limited capacity of the intestine to digest and absorb a lower density feed can have detrimental effects as the bird eats based on metabolic demand. The limiting factor of this cycle is dependent on ensuring the feed transit time through the intestine is sufficient for absorption to take place. Feed costs are impacted by a significant number of factors; availability and quality being the top drivers in this respect. So with this in mind the focus to better understand the intestinal function and physiology has increased. In doing so we have identified limiting factors that need to be regulated to optimize intestinal function: namely, optimizing nutrient digestibility and bioavailability, optimizing the intestinal environment to promote digestion and to limit factors which would reduce nutrient uptake, notably a low-grade localized immune stimulation. The former two factors are highly dependent on intestinal function and feed quality; however, the latter is where the phytotherapeutic interventions may exhibit their optimal function, by reducing the levels of mucous produced by a localized immune response to intestinal inflammation, this giving rise to a mucous layer coats the absorptive surface of the intestine, which impairs nutrient absorption. Phytotherapeutic products that are used should exhibit the following main functions: antimicrobial, antioxidant, immune-boosting and anti-inflammatory.

One product range that has shown significant success in the field is the supplementation with oregano-based essential oils. The active ingredients in these products are carvacrol and thymol. A review of studies has shown that thymol, and more so carvacrol, possesses a wide range of biological activities, including antibacterial, antiviral, antioxidant, anti-inflammatory, modulating of immunity response and regulating of the gut microbial population (Gholami-Ahangaran *et al.*, 2022) and, in meat-type chickens, promote growth and influence feed utilization. Furthermore, these compounds can affect the behaviour of layers and influence egg composition and eggshell thickness (Ramirez *et al.*, 2021). The main action of this is through the stimulation to increase the secretion of hydrochloric acid and pepsin in the proventriculus (He *et al.,* 2017).

Below is a summary of flock performance from a single farm serviced by our practice, whereby supplementation was introduced with an oregano-based essential oil in the third crop following a heightened coccidiosis challenge identified in flock 92 (Table 12.1). In a previous crop (flock 91) the farm had a confirmed diagnosis of rotavirus that negatively impacted on intestinal health, which as noted had a significantly negative impact on the weighted feed conversion ratio.

A significant benefit has been seen using oregano-based products when chickens are diagnosed with *Histomoniasis meleagridis* (Blackhead) which gives rise to focal hepatic necrosis and formation of significant prominent caecal cores. Field experience has shown that when these products are used prophylactically in higher risk farms, the incidence of disease, which ultimately contributes to mortality and loss of production, can be dramatically reduced (Hafez and Hauck, 2006).

Due to their antioxidant activity, phytotherapeutic products are used commonly during stress periods, including heat stress conditions, around the time of stressful flock interventions such as feed ration change, vaccination, bird transport, and following other health challenges that are diagnosed.

12.5 Product Choice and Field Implementation: Respiratory Health

Phytotherapeutics are also proving beneficial where birds are challenged with respiratory pathogens. Although many of the national flock will be vaccinated against common viral and bacterial pathogens as a routine procedure, there will be times when the field challenge breaks through the protection offered by vaccines, or more commonly if the vaccination is not administered according to the manufacturer's recommendations. This may result in an adverse reaction in the respiratory tract, affecting either the upper respiratory tract (nasal passages and trachea) or the lower respiratory tract (tracheal bifurcation, lungs

Table 12.1. Flock performance on farm A (approximately 180,000 broilers per crop) showing improved crop performance with regards to daily weight gain, weighted feed conversion, European Performance Efficiency Factor (EPEF) and antibiotic usage following supplementation in flock 93.

Flock No.	Mortality (%)	Rejects (%)	Average weight (kg)	Average age (day)	Daily weight gain (g)	Weighted feed conversion	EPEF	Antibiotic usage (mg/PCU)
91	3.88	0.46	2.00	34.61	57.66	1.75	322	40.73
92	2.12	0.97	2.12	34.81	60.77	1.64	365	12.63
93	2.95	1.08	2.37	35.96	65.81	1.51	412	5.31

and airsacs). As with its efficacy in the intestinal tract to down modulate the localized immune response and mucous production, similar effects are noted when used to target the respiratory tract. The main active ingredient is an essential oil derived from eucalyptus called cineole, which has shown to exhibit the following positive effects, namely, anti-spasmodic properties, expectorant and secretolytic properties and antimicrobial properties (Barbour *et al.*, 2012). The expectorant property is where these products working locally in the airways show their benefit, with assisting to clear the airways and facilitate breathing when the airway or lung is compromised, as well as improve the ciliary beat frequency of the mucous membranes (Abbasnia *et al.*, 2020). Their beneficial impact has been noted in field conditions where the following diseases have been diagnosed; *E. coli* air sacculitis, *Mycoplasma gallisepticum*, infectious bronchitis, avian metapneumovirus, infectious laryngotracheitis, *Ornithobacterium rhinotracheale* and fungal aspergillosis.

12.6 Conclusion

Putting together a specific programme for phytotherapeutic supplementation, whether it be as an in-feed or drinking water administration, relies on a good understanding of the farm facilities and compliance with the farm manager to carry out the supplementation effectively and to give accurate feedback on the flock performance. Prophylactic intervention has shown to be beneficial compared to implementation following diagnosis. The decision to stop the supplementation strategy will depend on the presenting and improving clinical signs, a review of the flock performance and taking into consideration the cost–benefit associated with the supplementation strategy.

References

Abbasnia, M., Mosleh, N., Dadras, H., Rezaeianzadeh, G. and Boroomand, Z. (2020) Effect of different herbal preparations on experimental viral respiratory complex of broilers: clinical, pathological and ciliary activity aspects. *Journal of Herbmed Pharmacology* 9, 277–285 doi: 10.1080/17450390600884500.

Barbour, E.K., Saade, M.F., Nour, A., Kayali, G., Kidess, S. *et al.* (2012) Evaluation of essential oils in the treatment of broilers co-infected with multiple respiratory etiologic agents. *The Journal of Applied Research in Veterinary Medicine* 9, 317–323.

Gholami-Ahangaran, M., Ahmadi-Dastgerdi, A.,Azizi, S., Basiratpour, A., Zokaei, M. and Derakhshan, M. (2022) Thymol and carvacrol supplementation in poultry health and performance. *Veterinary Medicine and Science* 8, 267–288. doi: 10.1002/vms3.663.

Hafez, M.H. and Hauck, R. (2006) Efficacy of a herbal product against *Histomonas meleagridis* after experimental infection of turkey poults. *Archives of Animal Nutrition* 60, 436–442. doi: 10.1080/17450390600884500.

He, X., Hao, D., Liu, C., Zhang, X., Xu, D. *et al.* (2017) Effect of supplemental oregano essential oils in diets on production performance and relatively intestinal parameters of laying hens. *American Journal of Molecular Biology* 7, 73–85.

Jin, S., Corless, A. and Sell, J. (1998) Digestive system development in post-hatch poultry. *World's Poultry Science Journal* 54, 335–345. doi: 10.1079/WPS19980023

Kogut, M.H. (2019) The effect of microbiome modulation on the intestinal health of poultry. *Animal Feed Science and Technology* 250, 32–40.

Ramirez, S.Y., Peñuela-Sierra, L.M. and Ospina, M.A. (2021) Effects of oregano (*Lippia origanoides*) essential oil supplementation on the performance, egg quality, and intestinal morphometry of Isa Brown laying hens. *Veterinary World* 14, 595–602.

CHAPTER 13
Biomarkers for Intestinal Health

Richard Ducatelle*, Evy Goossens, Venessa Eeckhaut and Filip Van Immerseel

Livestock Gut Health Team (LIGHT), Faculty of Veterinary Medicine, Ghent University, Belgium

Abstract

Poor intestinal health is one of the most important issues in the poultry industry of the last two decades. 'Overfeeding' with highly nutrient dense diets is the main trigger of these intestinal health issues, together with coccidiosis and some other common gut pathogens. All these pathologic conditions lead to upregulation of inflammation in the intestinal mucosa. The consequence of this is leakage of body fluids into the intestinal lumen and passage of intestinal content, including bacteria, from the intestinal lumen into the host tissues and blood vessels and from there directly to the liver. This may elicit an acute phase response in the liver. Additionally, the digestive system of high performing chicken breeds is susceptible to any defects in the quality of the feed. All the above may lead to dysbiosis, an ill-defined intestinal health disorder that is characterized by inflammation and by a shift in the intestinal microbiome. There is no simple laboratory test to diagnose dysbiosis because the etiology is ill defined. Even at necropsy, the diagnosis is not straightforward and thus the diagnosis usually is made (too) late in the disease process. This is the main reason why people are exploring the possibility to use biomarkers for early warning, allowing timely strategic interventions. Biomarkers are naturally occurring biological molecules found in blood, other body fluids, body secretions or tissues that are a sign of a physiological or pathological process/condition.

We used an experimental gut inflammation model and looked for dysbiosis-associated host proteins in intestinal content and faeces by proteome analysis to identify proteins suitable as biomarkers for dysbiosis. Other experimental models, including a necrotic enteritis model and a coccidiosis model, were

*Corresponding author: Richard.Ducatelle@UGent.be

DOI: 10.1079/9781800622746.0013

used to validate the selected markers. We classified the proteins discriminating between challenged and healthy birds into three categories: inflammation markers, leakage markers and epithelial cell differentiation markers. The first two are increased in challenged birds, while the third category is decreased. One particular biomarker stood out because it was an indicator of systemic inflammation leaking into the intestinal lumen and faeces: ovotransferrin. We and other investigators also looked for gut microbial biomarkers of dysbiosis, with promising results. All these results indicate that biomarkers will play a role in the monitoring of intestinal health in poultry. Nevertheless, the challenge will be in the interpretation of data since intestinal health is not synonym with optimal performance.

13.1 Introduction

Poor intestinal health is one of the most important issues in the poultry industry of the last two decades. This is related to the European Union (EU) wide ban on antimicrobial growth promoters in animal feed, which was introduced on 1 January 2006. It is also related to the continuous selection of both layer and broiler breeds for higher performance. The current level of egg-laying performance (and persistence) in layers and daily weight gain and feed conversion in broilers can only be achieved by massive feed intake, which puts excessive pressure on the physiology of digestion and absorption in the gastrointestinal tract of the birds. 'Overfeeding' with highly nutrient dense diets is the main trigger of intestinal health issues, together with coccidiosis and some other common gut pathogens. All these pathological conditions lead to upregulation of inflammation in the intestinal mucosa, characterized by infiltration of inflammatory cells and damage to the tight junctions that normally seal the intercellular space between the intestinal epithelial cells. The consequence of such processes is leakage of body fluids into the intestinal lumen and passage of intestinal content, including bacteria, from the intestinal lumen into the host tissues and on to the mucosal capillary lumina, which drain directly to the liver. The latter phenomenon is called bacterial translocation. Flooding of intestinal bacteria into the liver can be so severe as to elicit an acute phase response in the liver. This acute phase response is characterized by a shift in the protein secretion pattern of the liver and has a negative impact on the performance of the birds. The breeds of chickens that have been selected for high performance (either egg type or meat type) thus are susceptible to intestinal health issues, not only as a consequence of the quantity of feed ingested. Their digestive system is also susceptible to any defects in the quality of the feed, either owing to the presence of harmful substances, such as mycotoxins, the presence of poorly digestible ingredients, or myriad of other quality defects. Many of these conditions manifest as dysbiosis or dysbacteriosis.

13.2 Intestinal Health vs Dysbiosis, a Definition

The gastrointestinal tract is a specialized organ designed specifically for the digestion and absorption of nutrients, which is a truly challenging task. The first challenge is that ingested feedstuffs may contain potentially harmful substances. Indeed, when taking into consideration that edible materials, especially when of plant origin, not only contain nutrients but also contain potentially harmful and even toxic components (mostly secondary metabolites of the plant), it seems obvious that the digestive tract must be equipped with sophisticated machinery to selectively take up the valuable nutrients while at the same time blocking the uptake of toxins. The latter is taken care of by a family of receptors, generically named multidrug resistance proteins, with P-glycoprotein as the most important member. The selective uptake of nutrients in the small intestine is based on the expression by the absorptive villus epithelial cells of a large array of receptors for each of the different molecules that are required for the animal's metabolism. Thus, uptake of nutrients is a receptor-mediated active process through the transcellular pathway. The second challenge is the presence of numerous bacteria, derived from the oral cavity, the feed, the drinking water and other materials ingested from the environment. Considering that the intestinal content is characterized as a nutrient-rich, moist and warm environment, there is a constant risk of massive multiplication of potentially pathogenic bacteria. Moreover, the mere multiplication of apathogenic bacteria represents a threat, as they come in competition with the host for the uptake of the nutrients. In the healthy chicken, powerful antibacterial defence mechanisms are in place in the small intestine, in order to avoid competition for the nutrients, while mechanisms that support the growth of beneficial microbes are expressed in the caeca. Intestinal health thus can be defined as a balanced intestinal ecosystem, in which the host intestinal barrier and intestinal immune system are in balance with the feed and the bacterial populations that are present in the lumen of the different intestinal compartments. Even under these perfectly healthy conditions, repeated or even continuous slight damage to the intestinal mucosal surface is unavoidable, leading to what is now called 'physiologic inflammation' (Rabinowitz and Mayer, 2012).

Dysbiosis is a common, ill-defined intestinal health disorder in broilers that is accompanied by a shift in the intestinal microbiome. There is no simple laboratory test available to diagnose dysbiosis because the aetiology is not clearly defined. The diagnosis of dysbiosis in daily poultry practice therefore relies on sacrificing some birds and scoring the intestinal lesions at necropsy using standardized scoring systems (Teirlynck *et al.*, 2011). Numerous attempts have been made to try and find a biomarker for early and reliable diagnosis of dysbiosis and other intestinal health issues.

13.3 Biomarkers, a Definition

A biomarker is a naturally occurring biological molecule found in blood, other body fluids, body secretions or tissues that is a sign of a physiological or pathological process/condition. These signal molecules are used to diagnose a disease state or to follow up a metabolic process, including the response to treatment. Biomarkers are commonly used for early detection of disease states, to allow timely interventions, which may greatly enhance the success rate of treatments. This is typically the case in many types of cancer in humans, where the biomarkers are frequently used in population-wide screening programmes. In animals, biomarkers are most commonly used for the early diagnosis of disease states that constitute a diagnostic challenge, such as the intestinal health issues in intensive poultry production. The purpose of the present paper is to present an overview of candidate biomarkers for the evaluation and follow-up of chicken intestinal health. Such biomarkers can be derived from different sources, including blood, intestinal mucosal tissue, intestinal content or faeces.

13.4 Blood-derived Biomarkers of Intestinal Health

Severe inflammation in the gastrointestinal tract will trigger an acute phase response in the liver. The acute phase proteins produced by the liver are released in the circulation. They include ovotransferrin, alpha-1-antitrypsin and alpha-1-acid glycoprotein (Chen *et al.*, 2015; Gilani *et al.*, 2017; Rath *et al.*, 2009). Severe inflammations elsewhere in the body, however, can also trigger an acute phase response in the liver, making these blood-derived biomarkers non-specific.

Enzymes and other proteins secreted into the gut lumen may end up in the blood if there is severe leakage from the intestinal barrier. This is the case, for example, for fatty-acid-binding protein (Chen *et al.*, 2015) and diaminoxidase (Lei *et al.*, 2013). Citrulline is an amino acid produced mainly by enterocytes of the small intestine, which is released to some extent into the blood. It has been shown to be reduced in the blood of broilers in a nutritional dysbiosis model (Baxter *et al.*, 2019). Furthermore, metabolites and degradation products of the intestinal microbiota may end up in the blood following severe intestinal barrier damage. This has been reported for the microbial-derived D-lactate (Lei *et al.*, 2013) and for the lipopolysaccharide from Gram-negative bacteria (Chen *et al.*, 2015).

13.5 Intestinal and Faecal Biomarkers Derived from the Gut Microbiota

Microbiota-derived biomarkers include both changes in the microbial taxa and changes in the metabolites produced by the microbiota. Microbial signatures of dysbiosis and of other intestinal health issues in poultry have been reviewed recently by Bindari and Gerber (2022). Considering the paucity of information on this topic in poultry, however, it seems reasonable to use the human and lab-animal literature when attempting to find microbial biomarkers. Using this approach, the following suggestions can be made. There are reasons to expect that expansion of the phylum *Proteobacteria* (newest nomenclature: *Pseudomonadota*) (Shin *et al.*, 2015) and depletion of the butyrate producing families of *Lachnospiraceae* and *Ruminococcaceae* or specific genera within these families (*Faecalibacterium*, *Butyricoccus*), which have been reported as biomarkers for dysbiosis in humans, might also be suitable biomarkers of dysbiosis in poultry (for review, see Ducatelle *et al.*, 2018). *In vitro* research has shown that expansion of *Proteobacteria* and subsequent reduction in butyrate producers can be the consequence of an overexpansion of lactate producers (in humans mostly belonging to the genus *Bifidobacterium*) (Wang *et al.*, 2020). *Bifidobacterium* is of little significance in poultry. However, excessive expansion of *Lactobacillaceae* has been noted under conditions of intestinal stress and this may trigger that same cascade of *Proteobacteria* expansion and implosion of the major butyrate-producing families. Several human medical studies have identified shifts in specific genera or species of the faecal microbiome associated with specific disease states. These, however, cannot be expected to extrapolate easily to poultry.

Microbial metabolites that may be of value as biomarkers of intestinal health versus disease include short-chain and branched-chain fatty acids, indoles, hydrogen sulfide and other sulfur compounds, hydrogen, ammonia, etc. (Lee and Hase, 2014; Oliphant and Allen-Vercoe, 2019).

The use of shifts in the microbiota composition or shifts in microbial metabolites as biomarkers of intestinal health is hampered by the fact that these changes can be influenced strongly by the feed composition and by other environmental factors. Volatile metabolites constitute a sampling and analytical challenge.

13.6 Intestinal and Faecal Biomarkers Derived from the Host Mucosa

Host proteins appear in the intestinal contents not only by secretion, but also by the transmigration of cells such as heterophilic granulocytes, and by the sloughing of epithelial cells mostly from small intestinal villi. Several of these proteins are resistant to digestive enzymes and microbial degradation and can end up in the faeces. This has been the focus of intensive research efforts in recent years because such biomarkers seem to be less influenced by the feed formulation and are therefore expected to be more generic indicators of intestinal health. In 2018 we reported on the detection of the acute phase protein ovotransferrin in intestinal content and fresh faeces of broilers in experimental models of necrotic enteritis and coccidiosis (Goossens *et al.*, 2018). The fact that an acute phase protein, secreted in the general circulation by the liver, ends up in the intestinal content is a specific indication of intestinal leakage, which is an important characteristic of poor intestinal health. Barekatian *et al.* (2020) in their experimental models found alpha-1-antitrypsin, another acute phase protein, in the faeces, thus confirming the value of tracing acute phase proteins in faeces as indicators of severe gut barrier failure. Using different experimental models again, and focusing specifically on chronic intestinal inflammation, Dal Pont *et al.* (2021) detected a chicken calgranulin (MRP126), first identified by Bozzi and Nolan (2020) as an antibacterial protein from chicken heterophilic granulocytes, in faeces of experimentally challenged broilers. In that same study Dal Pont *et al.* (2021) also identified lipocalin-2 and hypoxia-inducible factor 1 as additional faecal biomarkers of intestinal inflammation.

We carried out an unbiased search for gut leakage biomarkers using a proteome analysis of ileal and colon contents from broilers exposed to intestinal stressors that are commonly encountered in the field (De Meyer *et al*., 2019). The proteome analysis was specifically focused on host proteins and allowed the identification of many proteins that were either increased or reduced in concentration in the intestinal content of challenged birds. Proteins that were reduced in concentration included aminopeptidase Ey and angiotensin converting enzyme, which are proteases secreted by differentiated epithelial cells of the small intestinal villi. Proteins that were increased in concentration included proteins released from damaged host cells such as histone H2A-IV and nucleophosmin, proteins from heterophilic granulocytes such as myeloid protein 1 and many more.

13.7 Conclusion

Today the most readily available samples for the quantification of biomarkers of intestinal health are blood and faeces. As mentioned above, a wide range of candidate biomarkers are available, derived from both types of samples. It is to be expected that additional new biomarkers will be identified in both matrices. In human blood samples, for instance, circulating mRNAs have been identified as candidate biomarkers of specific intestinal disease states (Tan *et al*., 2021). This is still a largely unexplored area in poultry research.

In faecal samples, candidate biomarkers of host origin are numerous and seem to represent the fastest way forward towards the development of bioassays with possible applications in the field. This process requires not only the development of the assay itself, but also the validation of representative sets of well-characterized field samples. In practical conditions, intensive broiler production is characterized by the use of highly nutrient dense feed formulas and high feed intake, putting considerable pressure on the physiology of the gastrointestinal tract. The best-performing broilers will thus inevitably present a certain degree of intestinal inflammation. The question will always be how far the system can be stretched before it collapses. The biomarkers will provide new tools to farm managers and veterinarians to closely follow up the flocks, but it will still be up to them to decide what level of risk they are prepared to take.

It can be expected that future developments in this field will not stop there. One can speculate that, in a further digitalized broiler industry, existing continuous monitoring systems, providing information on feed and water consumption, ambient temperature and humidity, and even external body temperature of the birds will be complemented with systems monitoring volatile molecules in the ambient air that come from exhaled intestinal microbial metabolites, such as H_2 and methane, and possibly other compounds, that could allow a continuous follow-up of the intestinal health of the birds.

References

Barekatian, R., Howarth, G., Wilson, N.L., Cadogan, D. and Wilkinson, S. (2020) Excreta biomarkers in response to different gut barrier dysfunction models and probiotic supplementation in broiler chickens. *PLoSOne* 15, 0237505.

Baxter, M., Latorre, J., Dridi, S., Merino-Guzman, R., Hernandez-Velasco, X. *et al.* (2019) Identification of serum biomarkers for intestinal integrity in a broiler chicken malabsorption model. *Frontiers in Veterinary Science* 6, 144.

Bindari, Y. and Gerber, P. (2022) Factors affecting the chicken gastrointestinal microbial composition and their association with gut health and productive performance. *Poultry Science* 101, 101612.

Bozzi, A. and Nolan, E. (2020) Avian MRP126 restricts microbial growth through Ca(II)-dependent Zn(II) sequestration. *Biochemistry* 59, 802–817.

Chen, J., Tellez, G., Richard, J. and Escobar, J. (2015) Identification of potential biomarkers for gut barrier failure in broiler chickens. *Frontiers in Veterinary Science* 2, 14.

Dal Pont, G., Belote, B., Bortoluzzi, C., Eyng, C., Sevastiyanova, M. *et al.* (2021) Novel models for chronic intestinal inflammation in chickens: intestinal inflammation pattern and biomarkers. *Frontiers in Immunology* 12, 676628.

De Meyer, F., Eeckhaut, V., Ducatelle, R., Dhaenens, M., Daled, S. *et al.* (2019) Host intestinal biomarker identification in a gut leakage mode in broilers. *Veterinary Research* 50, 46.

Ducatelle, R., Goossens, E., De Meyer, F., Eeckhaut, V., Antonissen, G. *et al.* (2018) Biomarkers for monitoring intestinal health in poultry: present status and future perspectives. *Veterinary Research* 49, 43.

Gilani, S., Howarth, G., Kitessa, S., Tran, C., Forder, R. *et al.* (2017) New biomarkers for increased intestinal permeability induced by dextran sodium sulphate and fasting in chickens. *Journal of Animal Physiology and Animal Nutrition* 101, 237–245.

Goossens, E., Debyser, G., Callens, S., De Gussem, M., Dedeurwaerder, A. *et al.* (2018) Elevated faecal ovotransferrin concentrations are indicative for intestinal barrier failure in broiler chickens. *Veterinary Research* 49, 51.

Lee, W-J. and Hase, K. (2014) Gut microbiota-generated metabolites in animal health and disease. *Nature Chemical Biology* 10, 416–424.

Lei, K., Li, Y., Yu, D., Rajput, I. and Li, W. (2013) Influence of dietary inclusion of Bacillus licheniformis on laying performance, egg quality, anti-oxidant enzyme activities, and intestinal barrier function of laying hens. *Poultry Science* 92, 2389–2395.

Oliphant K. and Allen-Vercoe E. (2019) Macronutrient metabolism by the human gut microbiome: major fermentation by-products and their impact on host health. *Microbiome* 7, 91.

Rabinowitz, K. and Mayer, L. (2012) Working out mechanisms of controlled/physiologic inflammation in the GI tract. *Immunologic Research* 54, 14–24.

Rath, N., Anthony, N., Kanna, L., Huff, W., Huff, G. *et al.* (2009) Serum ovatransferrin as a biomarker of inflammatory diseases in chickens. *Poultry Science* 88, 2069–2074.

Shin, N., Woong Whon, T. and Bae, J. (2015) Proteobacteria: microbial signature of dysbiosis in gut microbiota. *Trends in Biotechnology* 33, 496–503.

Tan, I., Coutinho de Almeida, R., Modderman, R., Stachurska, A., Dekens, J. *et al.* (2021) Circulating miRNAs as potential biomarkers for celiac disease development. *Frontiers in Immunology* 12, 734763.

Teirlynck, E., De Gussem, M., Dewulf, J., Haesebrouck, F., Ducatelle, R. and Van Immerseel, F. (2011) Morphometric evaluation of "dysbacteriosis" in broilers. *Avian Pathology* 40, 139–144.

Wang, S., Rubio, L., Duncan, S., Donachie, G., Holtrop, G. *et al.* (2020) Pivotal roles for pH, lactate, and lactate-utilizing bacteria in the stability of a human colonic microbial ecosystem. *mSystems* 5, 00645-20.

Abstract 26827
Mitigating Clinical Necrotic Enteritis with Probiotic Bacillus licheniformis

W. Van der Veken* and V. Hautekiet
Huvepharma NV, Belgium

Introduction: Necrotic enteritis (NE) usually leads to poor welfare, reduced profitability and major losses on farm. To help manage the gastrointestinal challenge, multiple feed additives have been developed, including probiotics.

Aims and Objectives: A commercially available *Bacillus licheniformis*, B-Act®, was trialled to help mitigate induced NE. The probiotic product is based on a single strain of *B. licheniformis* DSM 28710, already known to support technical performance parameters such as higher final body weights and improved feed conversion ratios (FCRs) in poultry.

Material and Methods: A 42-day induced NE trial was conducted in broilers, with both health and production performance parameters recorded. All animals received the same amount of *Clostridium perfringens* on day 19, 20 and 21 (1 ml/bird, 1.0 x $10^{8\text{-}9}$ CFU *C. perfringens*/ml). The used *C. perfringens* strain was known to have caused NE in a commercial operation. Three treatment groups were evaluated: a negative control (basal diet), a probiotic group (1.6x10^{12} CFU *B. licheniformis* DSM 28710/tonne of feed, start until finish) and an antibiotic group. Animals in this latter group were treated with oxytetracycline hydrochloride (OXT) according to label recommendations, for three consecutive days after the NE challenge.

Results: Even under the induced NE challenge, weight gains of the probiotic and OXT groups were similar to each other and significantly higher compared to the control at the end of the study (2.06 and 2.03 kg vs 1.8 kg, respectively; $P<0.05$). Weight gain of the probiotic group was already significantly higher compared to the control on day 21 ($P<0.05$), indicating a potential benefit of the probiotic even before clinical establishment of NE. FCR values followed

*Corresponding author: wouter.vanderveken@huvepharma.com

 Pre and Probiotics for Poultry Gut Health (eds H. Masey O'Neill, E. Burton and D. Scholey)
DOI: 10.1079/9781800622746.appx

a similar pattern, with a significantly lower overall FCR for the probiotic and OXT groups compared to the control (1.744 and 1.666 vs 1.922, respectively, d0-42; $P<0.05$). From a health perspective, NE lesion scores and NE mortality were both significantly improved in the probiotic and OXT groups compared to the control (0.18 and 0.08 vs 0.68, respectively; 2.5 and 0.6% vs 7.8%, respectively; $P<0.05$).

Conclusion: This trial demonstrates that feeding probiotic *B. licheniformis* DSM 28710 significantly improved health and growth performance parameters of broilers under a NE challenge. The results achieved with prophylactically administered *B. licheniformis* DSM 28710 were comparable to those realized with the therapeutic antibiotic treatment. Continuous administration of the probiotic thus has the potential to be a useful and practical tool to mitigate NE in commercial broilers, even more so as benefits were already recorded before the NE challenge took hold properly.

ABSTRACT 26840
Supplementation of Microbial Muramidase can Improve Egg Production in High Performance Laying Hens

F. Khattak[1]*, J. Houdijk[1], O. Ribo[2], M. Vadiei[2], E. Canet-Martinez, P. Segobola and M. Umar Faruk[2]

[1]Monogastric Science Research Centre, SRUC, Edinburgh, UK; [2]DSM Nutritional Products, Wurmisweg 576, 4303 Kaiseraugst, Switzerland.

Introduction: Exogenous muramidase enzyme (Balancius®) produced by *Tricoderma reesei* has been proven to hydrolyse peptidoglycans (PGNs) from bacterial cell wall debris. Hydrolysis of PGNs with Balancius® has great potential to reduce the negative effect of PGNs in the gut of poultry, thus improving gastrointestinal functionality and performance.

Aims and Objectives: Assess the efficacy of Balancius® to improve growth and egg production performance of laying hens.

Material and Methods: A total of 945 Lohmann brown hens aged 22 weeks old were allocated to one of the three dietary treatments where Balancius® was added at 0 (T1), 150 (T2) and 300 (T3) g/tonne to a wheat, maize, barley and soyabean meal-based diet to achieve a muramidase content of 0, 150 and 300 mg/kg feed, respectively. Each treatment was replicated 21 times with 15 hens/cage in a randomized complete block design. An enriched colony cage system with a 0.24 m^2 floor area was used. Dietary treatments were fed for 84 consecutive days (12 weeks) in mash form. Both feed and water were available *ad libitum*. Body weight gain (BWG), egg production (EP), daily feed intake (DFI), egg mass (EM), egg weight (EW) and feed conversion ratio (FCR) were determined. The data were analysed using Analysis of Variance (GenStat 19).

*Corresponding author: farina.khattak@sruc.ac.uk

Results: The in-feed recovery of muramidase in T1, T2 and T3 was 0, 128 and 303 mg/kg feed, respectively. During 0 to 84 days on trial, birds fed Balancius® at 300 g/tonne of feed had significantly higher BWG (+8%), EP (+1.6%), EM (+2%) and FCR (-1.8%) compared to the T1 control ($P < 0.05$). DFI and EW had no significant differences ($P > 0.05$) when compared between treatments, with T2 was similar to T1 and T3. The overall mortality was low (0.52%) and was not treatment associated.

Conclusion: The study suggests that supplementing hens' diet with Balancius® at 300 g/tonne of feed can improve growth and egg production performance, probably due to efficient gut functionality.

Abstract 26855 *Competitive Exclusion as a Tool to Reduce* Salmonella *Carriage: Multi-analysis*

V. Demey* and E. Chevaux

Lallemand SAS, France

Introduction: Competitive exclusion (CE) was described in the early 1970s as a tool to control *Salmonella* outbreaks in poultry farms.

Aims and Objectives: The objective of this multi-analysis is to combine all published data on the effect that one commercial CE product (Aviguard®) (AV) has on *Salmonella* carriage in chickens.

Material and Methods: Twelve publications were identified (reporting results of 31 experiments) comparing *Salmonella* carriage in broilers either treated or not with AV. In all studies, animals were challenged with *Salmonella* spp. in early life (<10 days – dose ranging from 4 to 9 log CFU/bird). Additional parameters *Salmonella*-positive animals (n=7), Infection Factor (IF) and Protection Factor (PF) (n=4) were added to the database when available. Log CFU counts were analysed using a Mixed Model (SPSS 24.0). The number of *Salmonella*-positive birds was analysed using Logistic Regression, IF and PF data using non-parametric tests.

Results: Data showed a significantly lower *Salmonella* carriage in animals treated with AV. For challenged control (CC), an average of 5.5 log CFU/g caecal content was reported. This is 3 log unit higher than the AV treated animals (2.5 log CFU/g caecal content, $P<0.01$). The odds for CC animals to be *Salmonella* positive are 5.5 higher than for AV birds ($P<0.05$). IF was numerically lower for the AV-birds (IF-CC: 3.4 vs IF-AV: 1.3) and PF was 14 times higher (PF-CC=1 vs PF-AV: 14.5).

Conclusion: Results confirm that the studied CE product is effective in reducing the carriage of *Salmonella* and can be a valuable tool for the poultry industry.

*Corresponding author: echevaux@lallemand.com

ABSTRACT 26856
Field Evaluation of a Microbial Solutions Treatment Programme on Broiler Farms

H. Elliott, L. Hurdidge and V. Demey*

[1]*Lallemand Animal Nutrition, France*

Introduction: Probiotics have been documented to have positive effects on poultry performance. More recently the benefits of the application of a protective biofilm after disinfection of broiler houses has been studied.

Aims and Objectives: The objective was to evaluate under field conditions the effect a treatment programme, combining both microbial solutions, on broiler performance and welfare.

Material and Methods: The trial was carried out in six standard broiler sheds (one farm, three consecutive crops, Ross 308). Half the houses were left to the farms' standard practices (standard disinfection, no probiotics) (CTR), in the other three houses the treatment programme, combining the positive biofilm (mix of selected Bacilli and lactic acid bacteria (Lalfilm Pro)) and probiotic (*Pediococcus acidilactici* via drinking water (Bactocell)), was applied (PROG). Each batch lasted for 35 days. Growth performance and mortality were recorded (per house and per crop). Carcass scoring was performed at the slaughterhouse. Litter samples were taken at day 14 and 35 for dry matter analysis. Data were analysed using a General Lineair Model (SPSS 25).

Results: Average live body weight was significantly improved for the birds from PROG (+4.6%, $P<0.05$). Dry matter of the litter of PROG houses was significantly higher compared to litter from CTR houses (CTR: 62.59% vs PROG: 67.31%). The improved dry matter content was associated with a significantly lower percentage of birds showing signs of pododermatitis (CTR: 10.6% vs PROG:2.6%, $p<0.05$).

Conclusion: The treatment programme resulted in improved performances of the birds, as well as a better litter quality, resulting in improved animal welfare.

*Corresponding author: vdemey@lallemand.com

ABSTRACT 26860
Identifying Differences in Tibial Ash, Morphometric Parameters, and Mineral Profile in 21-day-old Broiler Chicks of Varying Bodyweights

C. L. Elvis-Chikwem[1]*, E. Burton[2], G. White[1] and C. O'Shea[3]

[1]*School of Bioscience, University of Nottingham, UK;* [2]*School of Animal, Rural and Environmental Sciences, Nottingham Trent University, UK;* [3]*Department of Bioveterinary and Microbial Sciences, Technological University of the Shannon, Ireland*

Introduction: In poultry, lack of uniformity constitutes challenges from different viewpoints including growth, health and welfare. Significant variation in growth was observed in a recent broiler study, even with the same genetic line and management. This variation in growth comprises underweight chicks which fail to reach production targets in the early life. Understanding differences in the biological characteristics between the high- and low-weight chicks may provide useful insight into precise strategies to improve uniformity and growth of the underweight population.

Aims and Objectives: This study aimed at evaluating the differences in tibial ash, morphometric parameters and mineral profile of high- and low-weight chicks at 21 days of age. The objective was to obtain insight into drivers of variation in growth and to provide precise solution to the underweight.

Material and Methods: Day old chicks (male Ross 308) were raised in a deep litter pen and fed commercial diet. Ethical approval was granted by Nottingham Trent and University of Nottingham Ethics Committees. Tibial bones were extracted from the high weight (HW) (n=11; Average BW 1020 g) and low

*Corresponding author: chinwendu.elvis-chikwem@nottingham.ac.uk

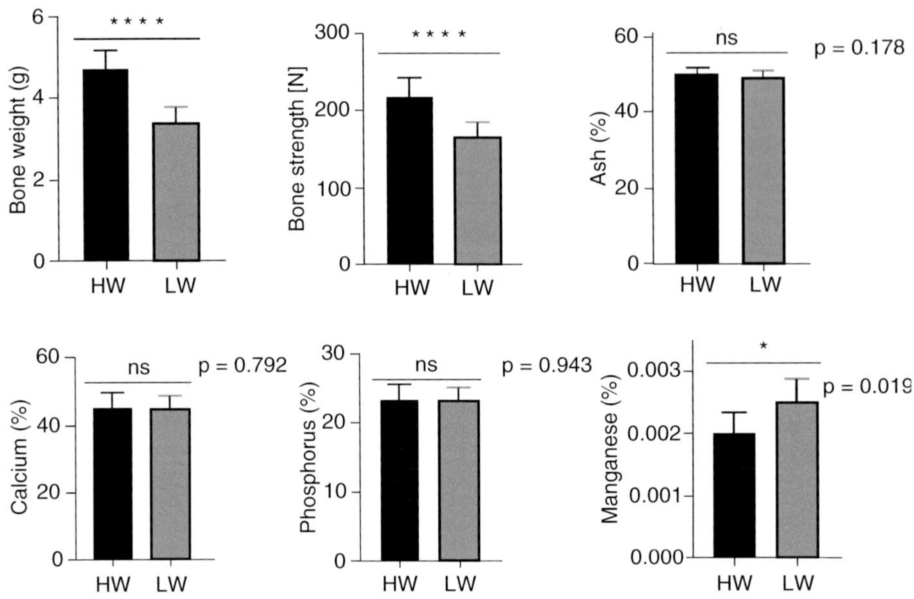

Fig. 1. Tibial bone measurements, ash, and mineral profile of HW and LW broiler chicks at D21.

weight (LW) (n=11; Average BW 746 g) broilers. Tibial weights (g) were determined using a precision balance and bone breaking strength was analysed with texture analyser (TA. XT plus 100). Tibias were oven-dried at 105^{0}C for 24 h and ashed at 600°C overnight to obtain ash content. Mineral profile was analysed by ICPMS, and statistics was performed using the student t-test of GraphPad prism (9.0) and significance was determined at $p<0.05$.

Results: The bone measurements and mineral profile are shown in Fig. 1. The HW group exhibited significantly greater bone weight and strength, compared to the LW group. No significant differences were observed in bone ash, calcium and phosphorus contents in both groups. However, manganese content was significantly higher in LW relative to HW group.

Conclusion: The HW group had greater bone weight and strength; this may indicate production of bone with optimal dimension for load support. No observed differences were apparent in ash or the two important bone minerals (Ca and P). The significance of the higher manganese content observed in the LW group is still unclear and requires further investigation.

ABSTRACT 26876
Gut Health Markers of the Efficacy of Probiotic Solution Bacillus subtilis *DSM 29784 in Broiler Chickens*

D. Preveraud*, S. Davies and F. Rudeaux

Adisseo, Heath by Nutrition, Antony, France

Introduction: Biomarkers to evaluate gut health are also difficult to determine with precise accuracy but they are necessary to predict animal performance and to test dietary strategies.

Aims and Objective: The aim of the present work was to evaluate the effect of the probiotic solution *Bacillus subtilis* DSM 29784 (Bs29784) on modulating these markers.

Material and Methods: In a first approach, birds were challenged with a diet high in non-starch polysaccharides (20% rye) and a selected group received Bs29784 on top. Gut scoring was performed on all birds; then, in a NE challenge, the Bs29784 group was compared to the control for histomorphological parameters. In another performance trial on healthy animals, microbiota in samples from cecal content was assessed in both control and Bs29784 groups. Finally, in a pilot study, key metabolites were measured in the intestine of birds receiving Bs29784.

Results: The macroscopical scoring method showed a negative correlation with villus length and positive correlation with T-cell infiltration. The Bacterial Enteritis score was 2.0 for the control group and decreased to 1.0 with Bs29784. On 42-day-old broilers, Bs29784 significantly increased villi length relative to control from 0.66 to 0.79 µm in the ileum, and from 1.25 to 1.46 µm in the caecum. The Bs29784 group showed higher alpha-diversity than control birds with the Chao and Shannon index. We also observed higher prevalence of *Ruminococcaceae* with the supplementation of Bs29784, especially for *Faecalibacterium prausnitzii* (+13.4%), that is correlated with villus

*Corresponding author: damien.preveraud@adisseo.com

length and CD3 area. We also observed that *Escherichia coli* at 21 days and *Clostridium perfringens* at 28 days were significantly reduced in the Bs29784 group. Finally, 13-day-old chicks showed that Bs29784 increased ($P < 0.01$) the intestinal content of nicotinic acid in ileum and jejunum, revealing a mode of action through which it can improve gut health.

Conclusion: Gut health biomarkers could be used for both research and practical purposes, and can be used to evaluate the efficacy of Bs29784 in broilers.

Abstract 26887
The Link Between Broiler Flock Heterogeneity and Caecal Microbiome Composition

R. Lundberg[1], C. Scharch[2] D. Sandvang[1] and A. Meuter[1*]

[1]Chr. Hansen A/S, Hoersholm, Denmark; [2]Feedtest, Wettin-Löbejün, Germany

Introduction: Despite low genetic variation of broilers and deployment of considerate management practices, there still exists considerable body weight (BW) heterogeneity within broiler flocks, which adversely affects the commercial value.

Aims and Objectives: The purpose of this study was to investigate the role of the caecal microbiome in weight differences between animals. Understanding how the gut microbiome may contribute to flock heterogeneity helps to pave the road for identifying methods to improve flock uniformity and performance.

Material and methods: 218 male broiler chicks were housed in the same pen, reared for 37 days, and at the end of the study the 25 birds with highest BW (Big) and the 25 birds with lowest BW (Small) were selected for microbiome analysis. Caecal contents were analysed by a hybrid metagenomic sequencing approach combining long and short read sequencing.

Results: We found that Big birds displayed higher microbial alpha diversity, higher microbiome uniformity (i.e. lower beta diversity within the group of Big birds), higher levels of short-chain fatty acid (SCFA)-producing and health-associated bacterial taxa such as *Lachnospiraceae*, *Faecalibacterium*, *Butyricicoccus* and *Christensenellales*, and lower levels of *Akkermansia muciniphila* and *Escherichia coli* as compared to Small birds.

Conclusions: Caecal microbiome characteristics could be linked to the size of broiler chickens. Differences in alpha diversity, beta diversity and taxa abundances all seem to be directly associated with growth differences observed in an otherwise similar broiler flock.

*Corresponding author: franme@chr-hansen.com

Abstract 27029
Probiotics Strains are More Resistant than Enterococcus cecorum *to the Antimicrobial Action of Botanicals*

M. Felici[1], B. Tugnoli[2], A. Doherty[2*], F. Ferrero[1], A. Piva[1,2] and E. Grilli[1,3]

[1]*DIMEVET, University of Bologna, Italy;* [2]*Vetagro S.p.A., Reggio Emilia, Italy;* [3]*Vetagro Inc, Chicago, IL, USA*

Introduction: Botanicals are well known for their antimicrobial action to inhibit intestinal pathogens in livestock. However, it is not clear whether they could also affect the beneficial microflora or supplemented probiotics.

Aims and Objectives: This study aimed to investigate *in vitro* the inhibitory and bactericidal effect of various botanicals against four probiotic strains (*L. acidophilus, L. paracasei, L. rhamnosus* and *E. faecium*) in comparison with *E. cecorum* a chicken intestinal pathogen.

Material and Methods: *Lactobacillus acidophilus* (*LA*), *Enterococcus faecium* (*EF*), *Lactobacillus rhamnosus* (*LR*), *Lactobacillus paracasei* (*LP*) and *Enterococcus cecorum* (*EC*) were used for this study.

Vanillin, thymol, carvacrol, eugenol, α-pinene, limonene and cinnamaldehyde were tested using a microdilution method from 30 to 0.24 mM (two-fold dilutions) for 24 h. Minimal inhibitory concentration (MIC) was considered the lowest dose giving null absorbance. Samples with null absorbance were seeded on agar to assess minimal bactericidal concentrations (MBC) after 24 h.

Results: MIC values against the four probiotics were found for vanillin, α-pinene, limonene (30 mM), eugenol (15 mM), thymol, carvacrol (7.5 mM) and cinnamaldehyde (3.75 mM).

*Corresponding author: aidan.doherty@vetagro.com

Against *EC*, MIC values were found for α-pinene, limonene (30 mM), vanillin, eugenol, cinnamaldehyde (3.75 mM), thymol and carvacrol (1.87 mM).

Vanillin, thymol, carvacrol and eugenol killed all the bacteria, with MBC values at least 2 times higher for probiotics than for *EC*. Cinnamaldehyde was bactericidal only on *LA*, *EF* and *EC*.

Conclusion: Probiotics are generally more resistant to botanicals compared to *E. cecorum*, suggesting that selected botanicals at the right concentrations could be used to positively modulate chicken microflora, controlling pathogens without interfering with beneficial bacteria.

ABSTRACT 27032
Influence of Manganese Amino Acid Complex in Layer Diets on Shell Organic Matrix Composition, Eggshell Ultrastructure and Mechanical Property

C. Torres[1]*, L. Linares[1] and H. Zhang[2]

[1]Zinpro Corporation, MN, USA; [2]Feed Research Institute, Chinese Academy of Agricultural Sciences, Beijing, China

Introduction: Manganese (Mn) is vital in the metabolism of glycosaminoglycans (GAGs) and uronic acid (UA), key components of the eggshell organic matrix.

Aims and Objectives: To investigate the effect of Mn sources on eggshell quality.

Materials and Methods: 324 Hy-line Brown layers were randomly assigned into three dietary treatments with six replicates each. From 23 to 46 weeks, layers were fed a Control corn-soybean diet (21,95 mg Mn/kg) or the control diet supplemented with 40 mg/kg Mn as sulfate ($MnSO_4$) or 40 mg/kg Mn as amino acid complexed (MnAAC). Eggshell ultrastructure analysis, GAGs and UA content were measured at 34 and 46 weeks of age. Eggshell thickness and eggshell breaking strength were evaluated at the end of every 4 weeks. Data were analysed using a one-way ANOVA at $P<0.05$.

Results: Egg production (23 to 46 weeks) was greatest with MnAAC supplementation followed by $MnSO_4$ ($88.4\leq 90.8 \leq 92.9$ %, p=0.016 Control, $MnSO_4$, MnAAC, respectively). Mn supplementation increased uric acid ($3.72 \leq 4.06 < 4.67$ mg/g, 46 weeks, p= 0.007 Control, $MnSO_4$, MnAAC, respectively) and GAGs ($2.11 < 2.31 < 2.40$ mg/g, 46 weeks, p=0.06, Control, $MnSO_4$, MnAAC, respectively) in the eggshell membrane at 46 weeks of age. At this time, the MnAAC fed group had eggshells with decreased mammillary cone width and improved effective and mammillary layer thickness resulting in

*Corresponding author: ctorres@zinpro.com

greatest resistance to break compared with other treatments ($31.7 \leq 34.3 \leq 36.94$, $p=0.005$, Control, $MnSO_4$, MnAAC, respectively).

Conclusion: Dietary MnAAC improved eggshell quality (breaking strength) by modifying the eggshell membrane organic matrix and the eggshell ultrastructure.

Abstract 27049

Bacillus *sp. PB6 Improves Performance and Alters the Microbiome of Breeding Pheasants Under Commercial Conditions in Poland*

K. Vermeulen, P. Bukowska, E. N'Guetta and A. Wealleans*

Kemin Europa NV, Toekomstlaan 42, 2200 Herentals, Belgium

Introduction: The impact of probiotic supplementation on pheasant breeder performance and microbiome was assessed in pheasant breeders under commercial conditions. Across the trial, probiotic supplementation increased daily laying rate, increasing and prolonging peak lay, as well as improved microbiome uniformity amongst individual birds, improved Firmicutes:Proteobacteria ratio in the caecum and significantly altered the caecal microbial composition towards a healthier state.

Aims and Objective: The availability of microbiome sequencing technology is especially exciting in less studied poultry species, including pheasants. A study was therefore conducted on a commercial pheasant breeder farm to assess the impact of probiotic supplementation on performance and microbiome composition.

Materials and Methods: A 12-week study was conducted at commercial pheasant breeding and rearing facilities in Poland. Half the birds on the farm were supplemented with *Bacillus* sp. PB6 at 2 x 10^8 cfu/kg. Laying data and mortality were recorded daily. Six hens/treatment were sacrificed at the end of the laying period. DNA was extracted from ileal and caecal content and 16S rRNA amplicon sequencing methods were applied to all samples. Alpha and beta diversity measures were calculated, as were differences in relative abundance at the phylum and family level.

*Corresponding author: Alexandra.Wealleans@kemin.com

Results: Supplementation increased daily laying rate (+5.9%, $P<0.05$), speed to and height of peak lay, and reduced the decline from peak. Rooster mortality was substantially decreased. The caecal ratio of Firmicutes:Proteobacteria increased following supplementation (1.36 control, 2.35 PB6). Supplementation induced higher uniformity in the distribution of bacterial taxa; compared to the control, fewer individual extremes were observed. Supplementation also numerically increased relative abundances of *Lachnospiraceae* and *Ruminococcaceae* (explained by increased *Faecalibacterium*, which contains only a single species: *F. prausniztii*, an abundant butyrate-producer), among others.

Conclusion: Application of *Bacillus* sp. PB6 to the diets of breeding pheasants improved performance, probably due to alterations in microbiome structure. The study also provided novel insight into the microbiome of laying pheasants.

ABSTRACT 27050
Impact of Probiotic Supplementation on the Microbiome of Broiler Chickens Housed on Two Commercial Farms

K. Vermeulen, S.De. Smet, E. N'Guetta, N. Smeets[1], S. Kirwan and A. Wealleans*

Kemin Europa NV, Toekomstlaan 42, 2200 Herentals, Belgium

Introduction and Aims: Availability of whole microbiome sequencing technology allows deeper assessment of microbiome shifts following probiotic supplementation and may be useful in elucidating modes of action. This study aimed to identify a probiotic-associated 'microbiome fingerprint' in broilers across two commercial broiler farms.

Materials and Methods: Half the birds on each farm were supplemented with *Bacillus* sp. PB6 at $1x10^8$ CFU/l drinking water. Diet and housing differed between the farms. On day 10 and 28, 10 male birds were selected from each group for ileal and caecal content collection. DNA was extracted and 16S rRNA amplicon sequencing methods were applied to all samples. Alpha and beta-diversity were calculated, and differences in relative abundance at the phylum and family level were assessed. $P<0.05$ was considered significant.

Results: Ileal and caecal alpha-diversity was higher in supplemented birds at d10; ileal beta-diversity on both farms was also significantly different between treatments at d10 and d28. Population shifts with supplementation were consistent across farms. Supplementation increased relative ileal levels of *Lachnospiraceae* (g: *Blautia*), *Ruminococcaceae, Enterococcaceae,* and *Leuconostocaceae* (g: *Weisella*); relatively more species of the *Oscillospiraceae* (g: *Flavonifractor*), *Christenellaceae* and *Bacteroidaceae* were seen in supplemented caecal samples.

*Corresponding author: Alexandra.Wealleans@kemin.com

Conclusion: The study demonstrates the limitations of 16s analysis as, on a genus level, many groups (i.e. *Weisella*) contain both beneficial as well as opportunistic pathogens. Optimising specific qPCR to target specific markers could also support which changes are beneficial, neutral or detrimental (e.g. *Candidatus savagella, Blautia*) and could identify the most predictable probiotic 'fingerprint'.

ABSTRACT 27051
Probiotic Supplementation Reduces Mortality by 25% and Antibiotic Use by >40% in Commercially Reared Pheasants

A. Wealleans[1]*, J. Reckitt[2], D. King[3] and S. Kirwan[1]

[1]Kemin Europa NV, Belgium, [2]Reckitt Agri Services, Yorkshire, UK; [3]Sandhill Veterinary Practice, Yorkshire, UK

Introduction: Probiotics have proven effective in promoting the health of various poultry species, to date there has been limited data on the effect in game birds.

Aims and Objective: This study aimed to look at the effect of probiotics on the growth and health of pheasants.

Materials and Methods: The trial was conducted on a commercial rearing farm in Yorkshire. Birds were housed in pens, with access to indoor shelter and a covered range; 24,000 birds were fed control diets, and 42,000 birds were treated with probiotic *Bacillus* sp. PB6 at 2 x 10^8 cfu/kg feed. Twelve birds/pen were weighed at hatch and at catch (at 47±5 days). Cumulative mortality data were recorded per batch at week 1, week 2 and at catch. Medication use data per batch were recorded by the farmer, and data collated and averaged by treatment.

Results: At catch, control birds weighed on average 523 g with supplemented birds weighing 538 g. Relative mortality was reduced by 24.77% following supplementation (5.33% control, 4.01% PB6; $P<0.05$). Supplemented pheasants received 3.4 courses of antibiotic treatment, with 5 courses given to control birds. Number of days treated was also reduced (23 days control, 14 days PB6, −39.13%). The reduction in antibiotics covered all used types of antibiotics, including amoxicillin, tiamulin and lincomycin. Overall, fewer

*Corresponding author: Alexandra.Wealleans@kemin.com

and shorter courses of antibiotics resulted in a 51.87% decrease in total mg of active ingredient used.

Conclusion: Supplementation with probiotics may be a valuable strategy in antibiotic reduction for the game industry, improving performance and bird health.

ABSTRACT 27229
Towards Understanding the Stimbiotic Effect of Xylo-oligosaccharide (XOS)-based Prebiotics on the Caecal Microbiota

S.E. Amir[1]*, D.J. Boocock[2,3], C. Coveney[2,3], H.V. Masey O'Neill[4], M.R. Bedford[4] and E.J. Burton[1]

[1]School of Animal, Rural and Environmental Sciences, Nottingham Trent University, UK; [2]The John van Geest Cancer Research Centre, Nottingham Trent University, UK; [3]The Centre for Health, Ageing and Understanding Disease, Nottingham Trent University, UK; [4]AB Vista Feed Ingredients, Marlborough, UK

Introduction: Supplementation of broiler diets with xylo-oligosaccharides (XOS) often improves performance via increases in short-chain fatty acids (SCFAs) in the caecum in volumes beyond that expected from the amount of XOS added.

Aims and Objectives: This study investigated the effect of feeding Ross308 broilers diets supplemented with XOS on protein expression of caecal microbiota. In the genomes of *Bacteroidetes* there exists derivatives of a prototypic, *Starch Utilisation System* (*Sus*) or '*Sus-like systems*' for degradation of complex polysaccharides. XOS acts as a signalling molecule that stimulates upregulation of the proteins of these *Sus-like* systems. Hence the aim of this study is to investigate the proposed '*stimbiotic mechanism*' whereby XOS stimulates caecal microbiota to upregulate the proteins involved in the degradation of complex polysaccharides.

Materials and Methods: Bacterial cells from broilers fed a control (CON) or XOS-supplemented diet were extracted from the caecal contents. The bacterial pellet was prepared for proteomic analysis using Liquid Chromatography Mass Spectrometry. Protein Pilot search and generation of a spectral library

*Corresponding author: saba.amir@ntu.ac.uk

in PeakView (SWATH microapp) revealed 418 unique proteins for SWATH quantitation, of which 382 were quantified.

Results: Heatmap generation and hierarchical clustering showed that the samples clustered clearly between CON and XOS treatment groups. Twenty-nine proteins were significantly differentially expressed with 20 proteins significantly higher in the XOS group, of which nine proteins are part of the *Sus-like* system.

Conclusion: The significant increase in the *SusC-like* and other membrane proteins in microbiota of XOS fed birds indicates that XOS stimulated the *Bacteroidetes* to upregulate their xylan-degrading (*Sus-like*) system.

ABSTRACT 27227
Effect of Waste Stream Substrate on the Growth Rate of Larvae: a Natural Source of Antimicrobial Peptides for Poultry Gut Health

A. Nourhan, E. Burton, M. Naeem*, D. Scholey and A. Alkhtib

School of Animal, Rural and Environmental Sciences, Nottingham Trent University, UK.

Introduction: Insect meal production on sea waste substrates may be an environmentally viable rearing substrate. Insect larvae meal (ILM) is known to contain antimicrobial peptides (AMPs) that could enhance poultry gut health. The viability of ILM as a source of AMPs is depending on establishing economically and environmentally production.

Aims and Objectives: The study examined the effect of differing waste stream substrates on the growth rate of larvae as a gut health enhancer for poultry.

Materials and Methods: Mealworm larvae boxes (n=8) were reared on three waste stream substrates; food waste, sausage waste and sea waste. Larvae biomass and substrate consumption were used to calculate the mean gain to substrate ratio alongside protein level.

Results: Sea waste substrate increased ($P<0.05$) mealworm biomass gain:substrate (g/g) and crude protein (%) compared to food and sausage waste streams. However, the dry matter (%) was increased ($P<0.05$) on food waste (Table 1).

Conclusion: Sea waste appears a viable waste stream substrate to raise the larvae meal. The next step is to assess prebiotic effects.

*Corresponding author: naeem.naeem@ntu.ac.uk

Table 1. Effect of waste stream substrates on the growth rate of larvae as a gut health enhancer for poultry.

Parameters	Treatment (Mean ±SE)			P-value
	Food waste	Sausage waste	Sea waste	
D0-14				
Mealworm biomass gain (g)	16.06 (0.171)^{c}	29.63 (0.509)^{a}	24.73 (0.461)^{b}	<0.001
Substrate consumption (g)	5.705 (0.416)^{a}	5.618 (0.371)^{a}	3.2 (0.667)^{b}	0.003
Mealworm biomass gain: substrate (g/g)	2.927 (0.229)^{c}	5.469 (0.437)^{b}	9.233 (1.102)^{a}	<0.001
Mortality by weight (%)	136.3 (10.73)	167.3 (13.33)	171.8 (8.29)	0.066
Mealworm dry matter (%)	58.17 (7.075)^{a}	44.25 (0.492)^{b}	40.57 (0.512)^{b}	0.005
Mealworm crude protein (%)	57.22 (0.661)^{b}	55.45 (0.253)^{c}	63.52 (0.127)^{a}	<0.001

$^{a-c}$ Means with differing superscripts within a row differ significantly ($P \leq 0.05$).

Abstract 27226
Measuring the Impact of Dietary Supplementation with Citrus or Cucumber Extract on Chicken Gut Microbiota using 16s rRNA Sequencing

F. Riva[1,2*], D. McGuinness[1], D. Mckeegan[1], J. Peinado-Izaguerri[1,3], G. Bruggeman[4], D. Hermans[4], M. Mclaughlin[1], D. Eckersall[1] and M. Bain[1]

[1]University of Glasgow, UK; [2]University of Zagreb, Croatia; [3]University of Veterinary Medicine and Pharmacy in Kosice, Slovakia; [4]Nutrition Sciences, Ghent, Belgium

Introduction: The need to find good alternatives to antibiotic growth promoters is a major challenge for the broiler industry. The presence of active molecules such as pectin, limonene, carotenoids and polyphenols in novel plant extracts, citrus and cucumber makes them interesting candidates.

Aims and Objective: The main aim is to test the hypothesis that the feed of broiler chickens, supplemented with citrus or cucumber extracts, has a positive effect on gut health through modification of the gut microbiota.

Materials and Methods: A dietary trial (ten birds x four replicates x three diets) was performed to test the hypothesis that a baseline diet (diet 1) supplemented with citrus (diet 2) or cucumber extract (diet 3) can modulate the gut microbiota. To study the microbiota profile, 12 birds per diet were culled at 14 and 28 days of age. The DNA from the jejunum and caeca was extracted and amplified by a 16s specific-PCR, sequenced using Illumina Sequencer and analysed with QIIME™.

Results: Alpha and beta diversity analysis showed that diets 2 and 3 did not significantly affect the bacterial diversity or taxonomy. However, linear

*Corresponding author: francesca.riva@glasgow.ac.uk

discriminant analysis effect size (LEfSe) showed that, in the jejunum, diet 2 increased the abundance of beneficial bacteria such as *Lactobacillus* and decreased *Enterococcus* and *Clostridium* at both 14 and 28 days of age. In the caeca, diet 2 increased the abundance of *Blautia* (another potentially beneficial bacterium) and decreased *Enterococcus* in each age group. Diet 3 had no major effects on the microbiota of the jejunum or caecum in either age group.

Conclusion: These results suggest that the use of citrus extract in broilers diets produces subtle beneficial effects on the microbiota, which could promote gut health.

INDEX

Page numbers in **bold** refer to figures, tables and graphs.